THE NEAT HOUSE GARDENS

To

JOAN THIRSK

for her advice and encouragement

THE NEAT HOUSE GARDENS

EARLY MARKET GARDENING AROUND LONDON

MALCOLM THICK

PROSPECT BOOKS
1998

Published by Prospect Books in 1998,
at Allaleigh House, Blackawton, Totnes, Devon TQ9 7DL.

British Library Cataloguing in Publication Data:
A CIP record for this book is available from the British Library.

Designed and set by Tom Jaine.

ISBN 0907325785

Printed by the Antony Rowe, Chippenham, Wiltshire.

Preface

Like, I suspect, many writers, I love research and am daunted by the task of writing. Consequently, this book has taken many years to complete but has involved many happy hours pouring over manuscripts, maps, drawings and books. My original plan was to write a comprehensive history of market gardening around London before the railway age, but the material uncovered suggested something different. As a start, I decided to find more information on the Neat House gardeners, who were said by early-eighteenth-century writers to be the best vegetable producers in the land. The discovery of a wealth of information on the gardens in the Duke of Westminster's archives dictated I should concentrate on the Neat Houses to such an extent that they now form the subject of a large part of the book. The information, especially from the late seventeenth and early eighteenth centuries, allowed me to form a picture of the gardens as thriving businesses and also to imagine what life was like in this small corner of a suburban parish. I have tried also to give an outline of the history of market gardening around London as an introduction to my detailed area study. Eventually, I hope someone will build on local histories which have been written to produce a full history of commercial gardening around the capital.

Research such as mine depends heavily on access to manuscript records and on the skill and goodwill of archivists. My general thanks go to all those who run the libraries and archives offices I have used and particularly to the Duke of Westminster for permission to consult the Grosvenor Estate records at Westminster and Chester; the staff of the City of Westminster Archives Collection; the

Chester Records Office; the Greater London Records Office; the Guildhall Library; and the Public Records Office. Many friends and colleagues have given advice and encouragement. Thank you Joan Thirsk for reading and commenting on my first draft, gently reminding me that I must push on with the book and imparting the skills needed for such research. I also thank members of the London Food Group, the Agricultural History Society and a seminar group at the Roehampton Institute who sat through talks which dealt with part of the research and made helpful comments, Tom Jaine for sympathetic editing, and my spouse Jane Card for keeping an eye on my style and grammar, drawing my attention to the delightful reference to waterborne gardeners in Richard Steele's *Spectator* essay, and urging me to complete this work.

Harwell, January 1998.

Contents

Illustrations

TABLES

ACKNOWLEDGEMENTS

The drawings used as figures 14, 18, and 22 are by Philippa Stockley, based on original manuscripts.

The author is grateful for permission to use the following illustrations: figures 2, 4, 5, 6, 8, 9, 16, the Guildhall Library, London; figures 3, 4, 5, 6, 11, 27, the London Topographical Society; figures 3, 21, the British Museum; figure 7, Geographical Publications Ltd; figure 10, The Geographical Association; original manuscript material used as sources for figures 14, 18, 22, the Duke of Westminster; figures 17, 25, 26, the Bodleian Library, Oxford; figures 1, 23, 24, Westminster City Archives.

Introduction

Market gardening is the Cinderella of gardening history, or perhaps to call it the Ugly Sister would be more descriptive. Here are no careful designs to delight the eye, only an ever-present whiff of manure. Artists seldom set out to depict market gardens and any which were included in paintings or prints are incidental parts of the landscape. But commercial gardening was an important source of food for many people in England, especially before the development of railways and refrigeration. It is a subject which deserves to be studied seriously.

What follows is an outline of the history of market gardening around London before 1800; a consideration of the role of commercial vegetable production in stimulating kitchen gardening generally and its possible influence on agricultural practice; and a history of one small but highly significant area of market gardening near London, the Neat House gardens near the river bank at Westminster. These gardens are now entirely swallowed by buildings, but they occupied the bend in the river just to the west of Vauxhall Bridge Road and south of Warwick Way.

'Oh! the incredible profit by digging of ground,' was Thomas Fuller's reaction to market gardening around London in 1662.[1] The wonder expressed by many writers in the seventeenth and eighteenth centuries at this relatively novel way of producing food for sale has not, alas, gripped many historians in this century. Market gardening has not had the attention which other aspects of garden history have received in recent years.

1. Thomas Fuller, *Worthies of England*, 2nd ed., 1811, vol. II, p. 353.

Gardening history, as exemplified by the Garden History Society and its publications, has concentrated almost exclusively on formal gardens and the materials and plants supplied for them. This study of flowers, ornamental plants, trees, lawns, ornaments and design links garden history to art, architecture, literature and plant breeding. There is good reason for this, as these very gardens form the most visible remains today, supplemented by a wealth of contemporary description and depiction. By contrast, most early market gardens lie buried beneath the towns and cities which they once served.

At the same time, much writing on English agricultural history of the period has concerned itself with the question of variations in output and institutional changes in the countryside. More specifically, the search for the agricultural revolution in England has been well to the fore. This has been the history essentially of grain and livestock production, with an emphasis on the expansion of food output and the contribution of agriculture to general economic progress. Market gardening has difficulty in finding a place here; it is part of the alternative agriculture expertly described by Joan Thirsk in her recent book,[1] quietly expanding near large towns throughout the period and taken up by some farmers on all or part of their land at times when the demand for mainstream field crops was depressed. Market gardening's innovation lay in the adoption of hotbeds and glassware to produce crops without regard to season. From the sixteenth century, market gardening was characterized, as its name implies, by production for the market, not subsistence, using capitalist forms of enterprise such as rented land, much hired labour and, as the period progressed, growing amounts of soil additives and glassware. The market was all: access to market was a major determinant of location, and responsiveness to the whims of consumers maximized profits. In other words, market gardeners succeeded by innovating to meet demand. Gardeners were the least dependent of all agricultural producers on the soils they worked since by dunging,

1. Joan Thirsk, *Alternative Agriculture*, Oxford, 1997.

adding new soil, drainage or irrigation they could, and did, transform their soils to meet the needs of production.

Market gardening thus has more of the appearance of industrial production, or at least of small workshop production, than does agriculture during much of the period under discussion. Occupying a small area of land close to major towns and cities, the significant amounts of food grown in these gardens has been little recognized. It is a paradox that the very factors most recently put forward as the essential constituents of the English agricultural revolution: the development of capitalist, market-orientated agriculture by the mid-nineteenth century, were present from the start in market gardening.

The novelty of market gardening at first confused contemporary observers. Other seventeenth-century writers than Fuller were amazed by the way gardeners carried on their enterprises, producing high profits from small plots of land in ways very different from most husbandmen. Their ambiguous status is exemplified by the incorporation of London gardeners as a City company in 1605, the only 'agricultural' trade to be a livery company. The Company was involved in a long-running legal dispute in the early seventeenth century as to whether the Fulham farmer-gardeners, who produced vegetables and field crops in open fields, were gardeners under the Company's jurisdiction. The finding against the Company was not a logical conclusion drawn from the facts displayed, but an economic and political decision.

Eventually, as I argue below, some gentlemen and writers on gardening and agriculture took an interest in the techniques of commercial gardening because they discovered features that held lessons for them, in both their gardens and their fields. Whilst it would be foolish to thrust market gardening anywhere near the centre of debate on the progress of English agriculture between the late sixteenth and early nineteenth centuries, it deserves recognition as having a minor part in bringing about agricultural change.

I

MARKET GARDENING
AROUND LONDON

Figure 1. View of the gardens at Millbank, from a painting by G. Arnald, engraved by John Hall, 1807.

CHAPTER I

The growth of demand

'When every clerke eates artichokes and peason...'

When the grain harvests failed in the mid-1590s, London and much of England faced famine and dearth. The shortages hastened the acceptance by Londoners of more garden vegetables in their diet. They constituted a significant cause of the expansion of market gardening around the capital. W. G. Hoskins described the years 1595 to 1597 as 'the Great Famine [which] extended over nearly all Europe, lasting for some three years. In Hungary it was said that the Tartar women ate their own children. In Italy and Germany poor people ate whatever was edible—fungi, cats, dogs, and even snakes.' In Sweden, people were 'so weak and their bodies so swollen that innumerable...died.'[1]

England's worst year was 1596. Despite a 'temperate winter' which 'made mens hearts to leape for joy, and the Barnes, as it were, to enlarge themselves for the receipt of this promised plentie,' unseasonable weather, especially in harvest time, with 'never ceasing raine' and 'tempestuous winds which choake out the corne when it would have been shorne,' destroyed early hopes and left the corn 'utterlie rotted and corrupted'. The shock of dearth in England was

1. W. G. Hoskins 'Harvest fluctuations and English economic history, 1480–1619', *Agricultural History Review*, xvi, 1, 1968, p. 38; Gustaf Uttterstrom, 'Climatic Fluctuations and Population Problems in Early Modern History', *Scandinavian Economic History Review*, 3, Stockholm, 1955, pp. 27–28. This chapter is based on the author's articles, 'Roots and other garden vegetables in the diet of Londoners', *Oxford Symposium on Food & Cookery*, 1989, pp. 228–235, and '"Superior Vegetables", greens and roots in London, 1660–1750', in *Food, Culture & History*, ed. G. & V. Mars, 1993, pp. 132–151.

intensified by contrast with earlier abundance: the grain harvests of 1591, 1592 and 1593 were all good. These bountiful harvests led the government to repeal the Anti-Enclosure Act of 1563, and to allow limited export of wheat.[1]

The bad weather affected all major bread grains. Wheat, barley, oats and rye prices were very high, and dietary alternatives such as peas, beans and dairy produce also became much more expensive. Cattle and sheep prices did not rise above average but, as such meat was normally beyond the pockets of the poor, one would not expect a close relationship between meat and grain prices in years of dearth. That scarcity caused distress in London is clear from comments recorded at the time. Furthermore, analysis of burials in selected London parishes shows that mortality rose in 1597, following the bad harvest of the previous year. Unless some unrelated and undisclosed outbreak of disease occurred, people in London in 1597 died from hunger or from illnesses caused by malnutrition.[2]

Government response came from the Privy Council which issued instructions on action to be taken in times of dearth in proclamations and exhortations to provincial justices and councils. In order to ensure there was grain available for bread, the Council sought to restrict its non-food uses: malting, brewing, and starch-making. Exports were banned, imports encouraged and even hijacking of foreign grain ships in the Channel authorized. Lastly, the Council tried moral suasion, The clergy was instructed to preach against the gluttony and to encourage charity; to urge the poor to be patient and accept their lot. The Lord Mayor was ordered to prevent excessive feasting in the City of London.[3]

1. W. Barlow, trans., *Three Christian sermons made by Ludovike Lavatore, Minister of Zuricke in Helvetia, of Famine and Dearth of Victualls*, 1596; Henry Arthington, *Provision for the poore, now in penurie*, 1597; Hoskins, op. cit., pp. 37–8.
2. A.B. Appleby, 'Nutrition and Disease: The case of London, 1550–1750', *Journal of Interdisciplinary History* , VI, 1, 1975, p. 5; A.B. Appleby, *Famine in Tudor and Stuart England*, 1978, pp. 138–9.
3. *The Agrarian History of England & Wales*, vol. IV, 1967, pp. 581–2; Appleby, *Famine in Tudor and Stuart England*, pp. 140–5.

London, a city of some 200,000 inhabitants and by far the largest urban area in England, posed particular problems. The poor (and most of the rich) had no gardens and were dependent on food brought into the city. The Privy Council understood the problem of maintaining the flow of food despite shortages in areas which supplied the capital: 'wee have founde exceeding great difficulty to reconcile the wantes of the citty and countrie, the one requiring great supply, the other not so able in these as in other tymes to affoarde such stoare.' Nevertheless, the Council had frequently to order provincial authorities to release grain for London. In October 1596, Norfolk justices were told to hand over 1,520 quarters of grain to London bakers, 'because the cittie of London, being the chief place and resort of this realme, may in no wise be left to[o] much unprovided, and cannot be other wise sufficiently healped but by supplie out of other counties, amonge the which we suppose that the countie of Norfolk may best spare some reasonable porcion.' The London authorities themselves also tried to control the day-to-day supply and price of food in the City and suburbs and encouraged the rich to give food and alms to the poor.[1]

These measures, however, provided respite but not deliverance from dearth and its consequences. The situation also excited preachers, publicists and pamphleteers to propose their own remedies, divine or ingenious. One was Hugh Platt, who wrote *Sundrie new and Artificiall Remedies against Famine* in 1596. This explored various substitutes for conventional bread flours and alternatives for bread itself, for instance cakes made of parsnip meal.[2]

Another pamphlet did not appear until 1599 and may be thought a reasoned response to the years of famine rather than an immediate solution. It does, however, shed light on the effect of harvest failures on the diet of the poor. Entitled *Profitable instructions for the*

1. *Acts of the Privy Council*, 1596, p. 269; 1597–98, pp. 291–92.
2. Hugh Platt, *Sundrie new and Artificiall Remedies against Famine, Written by H. P. Esq. uppon thoccasion of this present Dearth*, 1596

manuring, sowing and planting of kitchen gardens. Very profitable for the common wealth and greatly for the helpe and comfort of poore people, its author was Richard Gardiner, a wealthy and philanthropic dyer of Shrewsbury who also ran a market garden of at least four acres and drew on his considerable experience of gardening to write a 'short and simple penning of this my practise and experience in Gardening'. Gardiner, in a preface 'to his loving neighbours and friends, within the towne of Shrewsburie' explains that in his 'olde age, or last daies,' he 'would willingly take my last farewell with some good instructions to pleasure the general number'.[1]

The work is extremely practical. The first pages provide advice on vegetable growing equal to many modern gardening manuals. He explains how to raise and save the seeds of carrots, cabbages, parsnips, turnips, lettuce, beans, onions, cucumbers, artichokes, radishes, porrets and leeks. This section may have been written before the scarcities of the 1590s, possibly as part of a larger projected book, for only in the concluding sentence is the dearth noticed when he hastily excuses further general discussion of vegetables with: 'I could yet heerein take occasion to write of divers rootes and hearbs, for sallets, to bee planted and sowed in gardens, which do not serve my purpose, for I rather desire to provide sufficient victuals for the poor and greatest number of people, to relieve their hungrie stomackes, then to picke dainty sallets, to provok appetite to those that doe live in excesse, the which God amend.' He then begins what is almost a second pamphlet, specifically on the famine years, vividly describing his part in helping the poor 'in the great dearth and scarcitie last past in the Countie of Salop and elsewhere, for with lesse garden ground then foure ackers planted with Carrets, and above seaven hundreth close cabbedges, there were many hundreds of people well refreshed thereby, for the space of twenty daies, when

1. 'Richard Gardiner's "Profitable Instructions", 1603, ed. Dr. Calvert', *Shropshire Archaeological and Natural History Society*, Series II, vol. 4, Shrewsbury, 1892, pp. 241–2; Richard Gardiner, *Profitable instructions for the manuring, sowing and planting of kitchen gardens*, 1599.

bread was wanting amongst the poore in the pinch or fewe daies before harvest. And many of the poore said to me they had nothing to eate but onely carrets and Cabedges, which they had of me for many daies, and but onelie water to drinke. They had commonly sixe ware poundes of small close Cabedges for a penny to the poore. And in this manner did I serve them, and they were wonderfull glad to have them, most humbly praising God for them all.'

He sold cheaply to the poor during the famine. In normal times large close-cabbages were 2 lb for a penny and small ones 5 lb for a penny. Other vegetables he usually sold as follows: large artichokes, 1d each; small artichokes, 1d for two or three; green beans, 1d a quart; large turnips, 2d a stone; yellow carrots, 2d a stone, 3d a stone from January to March.

Gardiner thought carrots the most important vegetable for the relief of hunger and gave a number of recipes for them, noting that, 'this last dearth and scarcitie hath somewhat urged the people to proove many waies for their better reliefe whereby I hope the benefit of Carret rootes are profitable'. He begins with the 'use of them amongst the better sort by the Cookes' [i.e. carrots for the rich], recommending that carrots be cut up and boiled to season broth, boiled with powdered [i.e. salt] beef and pork and with any meat the poor can afford. Red carrots make 'daintie sallets' with vinegar and pepper, to go with roast meat. Mixed into potages of any kind 'they effectually make those pottage good, for the use of the common sort....Carrets well boyled and buttered is a good dish for hungrie or good stomackes.'

Addressing himself specifically to dearth he wrote, 'Carrets in necessitie and dearth, are eaten of the poore people, after they be well boyled, instead of bread and meate. Many people will eate Carrets raw, and doe digest well in hungry stomackes: they give good nourishment to all people, and not hurtful to any, whatsoever infirmities they be diseased of, as by experience doth proove by many to be true.'

The people of London made what provision they could to replace bread in their diet. Both Platt and Gardener recognized the potential of roots and other garden vegetables to feed the poor. In this they reflected increased consumption of these foods in the late sixteenth century. The expansion of market gardening around London was itself a response. In 1577, William Harrison found the poor eating, 'melons, pompions, gourds, cucumbers, radishes, skirrets, parsneps, carrels, cabbages, naevewes, turneps, and all kinds of salad herbes'. Later, Thomas Cogan remarked of parsnips and carrots, 'the rootes are used to be eaten of both first sodden, then buttered, but especially Parseneppes: for they are common meate among the common people all the time of Autumne, and chiefly uppon fish daies.'[1]

That roots were available in the markets was clear, although quality could not invariably be assured. There was a proposal in 1593 'for restraint of those that let out cellars and sheds under stalls where herbs, roots, fruits, bread, and victuals are noisomely kept till they be stale and unwholesome for man's body, and then, mingled with fresh wares of the same kind, are brought forth into the markets and there sold to the great deceit and hurt of the people.'[2]

In the terrible dearth of the 1590s, root vegetables played a larger part than before in feeding London's poor. As well as local produce, London received roots, mainly carrots, from the East Anglian ports of Norwich, Yarmouth, and Colchester. Most were shipped through Yarmouth: between October and March 1593–4, 281 tons and 600 bushels went to London; in 1597–8, 600 tons and 600 bushels were sent; and in 1598–9, 639 tons and 1 last. These totals may be small in comparison with the 111,075 quarters of grain which came to the port of London in the seven months ended May 26th 1597, but would have been a welcome addition to total food supplies.[3]

1. William Harrison, *The Description of England*, ed. Georges Edden, New York, 1968, pp. 263–4; Thomas Cogan, *Haven of Health*, 1596, p. 63.
2. British Library Lansdowne MS 74, ff. 75-76.
3. Appleby, op. cit., p. 139.

The suppliers from East Anglia were mainly Dutch and Flemish refugees who had settled at Norwich, Colchester, and Yarmouth after 1550 to escape religious persecution. They began market gardening and, in particular, the intensive production of root vegetables. At Norwich in 1575, they were said to 'digge and delve a grete quantitie of grounde for rootes which is a grete succor and sustenaunce for the pore'.[1] The crops' potential not only impressed shrewd observers such as Platt and Gardiner. Following the shock of near-famine in the 1590s, practical men saw profit in increasing the supply of root vegetables to London markets. In about 1600, many Dutch gardeners moved to the Surrey bank of the Thames and, fifty years later, old men in Surrey could still recall 'the first gardiners that came into those parts, to plant Cabages, Colleflowers, and to sowe Turneps, Carrets, and Parsnips, to sowe Raith [early ripe] Peas'.[2]

Across the Thames, husbandmen in suburban Middlesex turned to roots and other vegetables. John Norden in 1607 mentioned Fulham in a list of carrot-growing areas and by 1616 the husbandmen and gardeners there were in legal conflict with the recently formed Gardeners' Company of London because of their root production, which the Company sought to regulate. The dispute continued until 1633 by which time root growers in neighbouring Kensington and Chelsea were also involved. The Company was alarmed at the large scale of production undertaken by the Middlesex men and sought to bring them within its regulations on size of garden ground and numbers of employees. The Company lost the case, for the pragmatic reason that London could not afford restrictions on this source of food for the poor. These Middlesex producers, 'by this manner of husbandry and ymployment of their grounds [furnish] the Cittys of London Westminster and places adjacent ... with above fower and twenty Thousand loads yearly of Rootes.'[3]

1. *State Papers Domestic*, Elizabeth, vol. 20, no. 49.
2. Samuel Hartlib, *His Legacie of Husbandry*, 1655, pp. 8–9.
3. John Norden, *The Surveyor's Dialogue*, 1607, p. 207; Corporation of London, City Repertories, 33, f. 74, recto; 49, ff. 261–3.

Because the expansion of market gardening increased supply, vegetables were more commonly eaten in London after 1600. A London schoolboy's diet in the early seventeenth century reflects this change:

> Our breakfast in the morning, is, a little piece of bread not buttered, but with all the bran in it, and a little butter, or some friute, according to the season of the yeare. To dinner we have herbes, or everyone a messe of porridge. Sometimes turneppes, coleworts, wheat and barley in porridge, a kind of delicat meate made of fine wheat flower and eggs. Upon fishe dayes, fleeted milke, in deepe porrengers (whereout the butter is taken) with some bread put in it. Some fresh fishe, if in Fish street can be had at a reasonable price. If not, salt fish, well wattered. After pease, or fitches, or beans, or lupins.[1]

In a crude and satirical comment on London's eating habits and sanitary arrangements Ben Jonson wrote, in 1616, of a summer voyage up the Fleet Ditch, a minor tributary of the Thames between Westminster and the City,

> ...How dare,
> Your daintie nostrills (in so hot a season)
> When every clerke eates artichokes and peason,
> (Laxative lettus and such windie meate)
> Tempt such a passage? When each privies seate
> Is fill'd with buttock? And the walls doe sweate
> Urine and plaisters? When the noise doth beate
> Upon your ears, of discomforts so unsweet?[2]

It is possible that some people may have consumed too many vegetables, so that they developed hunger-oedema, a condition characterized by waterlogged tissues and swollen limbs: the 'moist and loose flesh' which a London apothecary noted as a consequence

1. G. E. Fussell, *The English Rural Labourer*, 1949, p. 29.
2. Ben Jonson, *Epigrammes* 133, 'On the famous voyage', 1616.

of turnip eating in 1629? But the increasing popularity of roots is no wonder: they were inexpensive, palatable, and easy to prepare.[1]

✻ ✻ ✻ ✻ ✻ ✻

So far the impetus given to vegetable consumption by scarcity has been emphasized, with the poor being forced to adapt their diet to include roots, cabbages and other bulky vegetables. After 1650, a century of stagnating or falling food prices, the result mainly of the interaction of a slower rise in population with increased agricultural productivity, allowed English men and women greater choice of diet and forced many farmers to grow alternative crops to make a living. In these altered circumstances the demand for vegetables and salads, especially in London, changed its character. The poor still ate their roots, but the rich (who had most choice of what to eat) developed a taste for vegetables as part of a new and fashionable diet. This was, as will be revealed, a sophisticated taste, which stimulated both further growth of gardening around the capital and the technical development of commercial horticulture.

By the early 1700s the upper and middle classes in London led the way in what has been called the 'Consumer Revolution' of the eighteenth century. This phenomenon marked a transformation in the attitude of people towards the acquisition of goods and services: 'more men and women then ever before in human history enjoyed the experience of acquiring material possessions'. Consumption was stimulated by fashion, a desire for novelty in everything, food, and more particularly vegetables, included.[2]

In this period, reliable contemporary statistics of domestic consumption are few and are, at best, informed guesses. However, two statistical pioneers, Gregory King in the 1680s and Jacob Vanderlindt in the 1730s, do include some data on fruit and vegetable consumption in their works. Vanderlindt estimated that food and

1. J. C. Drummond & A. Wilbraham, *The Englishman's Food,* 1939, pp. 106–9.
2. N. McKendrick, J. Brewer, J.H. Plumb, *The Birth of a Consumer Society*, 1983, p. 1. See also Carol Shammas, *The Pre-Industrial Consumer in England*, 1990.

drink cost a seven-member family in 'the middling station of life' £76 16s per annum in 1734, 33 per cent of its total expenditure. Gregory King divided his estimates into many income bands and his per capita figures for middle-class food consumption in 1695 range from £5 to £20 per annum, say £35–£140 for seven-member households. King costed fruit and vegetable consumption on a national per capita basis at 4s 4d per annum in 1695, 5.75 per cent of his estimated total per capita food bill. Applying this percentage to the total food bills for the middle classes outlined above, we arrive at annual expenditure on fruit and vegetables of £4 7s for Vanderlindt's household and between £1 19s 6d and £7 18s 1d for King's.[1]

These figures, modest but significant elements in household expenditure, sit oddly with a statement by Richard Bradley in 1726 that, 'It is not very rare to see Bills from Fruiterers and Herb-shops, of one Winter's standing, to amount to Sixty, Eighty, an Hundred, and sometimes an Hundred and Fifty Pounds, where the Families are large.' Bradley, author of works on botany, agriculture, gardening and cookery, Professor of Botany at Cambridge, and intimately acquainted with both commercial gardening around London and the marketing of vegetables in the capital, is an author whose opinions should not be dismissed lightly.[2]

It may be possible to reconcile Bradley's figures with those of the statisticians. Bradley includes upper-class as well as middle-class households; presumably the ostentatious consumption of the very rich might account for his highest estimates. Bradley was writing of London where the largest part of a rich household's consumption of vegetables would be purchased, rather than home-grown. He included fruit in his figures. Imported fruit was expensive, as were

1. Peter Earle, *The Making of the English Middle Class*, 1989, pp. 269–281; Gregory King, *Natural and political conclusions upon the state and condition of England*, edited by G.Chalmers, 1810, pp. 48–9, 67.
2. Richard Bradley, *A General Treatise of Husbandry and Gardening*, 1726, p.150; Richard Bradley, *Country Housewife and Lady's Director*, 1736 (reprinted 1980, introduction by Caroline Davidson), pp. 10–31.

exotics like grapes and pineapples produced in England in greenhouses; these would inflate the bills. But the most significant reason why Bradley may have been right was the willingness of the well-off in the early eighteenth century to consume vegetables which were expensive to produce and commanded high prices at market.

Detailed comments on food are not plentiful in diaries or literature of the period and such as there are tend to mention the main, meat ingredient of a meal, not the vegetables on the side. Some of the best evidence comes from foreign visitors. The oft-quoted passage in the memoirs of Henri Misson, dating from the 1690s, bears repeating:

> Generally speaking, the English Tables are not delicately serv'd. There are some Noblemen that have both French and English Cooks, and these eat much after the French manner; but among the middling Sort of People they have ten or twelve Sorts of common Meats, which infallibly take their Turns at their Tables, and two Dishes are their Dinners: a Pudding, for instance, and a Piece of Roast Beef; another time they will have a piece of Boil'd Beef, and then they salt it some days before hand, and besiege it with five or six Heaps of Cabbage, Carrots, Turnips, or some other Herbs or Roots, well pepper'd and salted, and swimming in Butter: A Leg of roast or boil'd Mutton, dish'd up with the same dainties, Fowls, Pigs, Ox Tripes, and Tongues, Rabbits, Pigeons, all well moistened with Butter, without larding: two of these Dishes, always serv'd up one after the other, make the usual Dinner of a Substantial Gentleman, or wealthy Citizen.[1]

Pehr Kalm, a Swede visiting London in 1748, also ate dinners whose main ingredient was butcher's meat; he noted, however, that 'they take turnips, potatoes, carrots, &c from the dish, and lay them in abundance on their plates'.[2] Kalm also mentions other vegetables

1. Earle, op. cit., pp. 276–7; *Englishmen at rest and play*, edited by R. Lennard, 1931, pp. 233–4.
2. Pehr Kalm, *Kalm's Account of his Visit to England on his Way to America in 1748*, edited by J. Lucas, 1892, pp. 7, 14.

commonly eaten; lettuce, salad sprouts, cabbage, spinach, turnips, and green peas.

These are observations of the day-to-day meals of middle-class Londoners; Misson also commented on the French cooks employed by the nobility. French influences were not new; in the first decade of the seventeenth century, Sir Thomas Overbury characterized a 'French cook in England' as one who produced salads, vegetables, mushrooms, and sauces in abundance but little meat and small portions. He fed not 'the Belly, but the palate....He dares not for his life come among the butchers, for sure they would quarter and bake him after the English fashion, he's such an enemy to beef and mutton.' Despite such criticism, a French cook was a mark of status; Pepys recorded in 1660 how Lord Mountagu, finding his fortune much improved, 'did talk very high how he would have a French Cook and a Master of his Horse'.[1]

Many fought a rearguard action against French cookery into the eighteenth century. Richard Steele lamented the demise of roast beef and plain cooking in 1710 and a poet satirically regarded French cuisine so essential that the lack of it at Public Schools and Universities interrupted 'Progress in Learning'. But by 1750 the ubiquity of ragouts, omelettes, fricassees, and other made-dishes in cookery books indicates that French cooking had made a marked impression in the kitchens of the better-off in England, despite the continuation throughout the period of a tradition of more robust and rural recipes in middle- and even upper-class cooking, which tempered the influence of French cuisine on all but a small circle of the richest London households. Hannah Glasse expressed the ambiguous attitude to French cookery at the time when she condemned 'the blind Folly of this Age, that they would rather be impos'd on by a French Booby, than give Encouragement to a good English cook,' but nevertheless gave English cooks many French-influenced recipes to

1. H. Morley, *Character Writings of the Seventeenth Century*, 1891, pp. 84–5 Samuel Pepys, *Everybody's Pepys*, edited by O.F. Morshead, 1955, p. 55.

try. French cuisine, with its imaginative use of vegetables and smaller emphasis on meat, must have contributed something to the increased demand in London for garden produce, especially high-quality crops which were expensive to cultivate.[1]

The increasing taste for gardens, a trend clearly discernible after the Restoration in much of society, fostered more interest in vegetables and fruit amongst the middle and upper classes. In the 1660s John Worlidge, a reliable observer, found 'scarce a cottage in most of the southern parts of England, but hath its proportionate garden, so great a delight do most men take in it,' and John Aubrey thought that 'in the time of Charles II gardening was much improved and become common'. The Earls of Bedford, who hired an experienced gardener to improve their kitchen gardens at Woburn after 1671, represented a general trend amongst the nobility.[2]

Writers justified producing ever more books on gardening for the gentry by pointing to its expansion. John Laurence wrote in 1717 of 'Gardening being of late Years become the general Delight and Entertainment of the Nobility and Gentry, as well as the Clergy of this Nation'. Two years later, George London and Henry Wise thought another book necessary because, 'Of late years...Gardening and Planting have been in so great Esteem'. An early historian of horticulture looked back in 1829 on gradual improvements in the seventeenth century which 'burst forth in splendour' in the eighteenth.[3] Research has confirmed that there was a steady, if small,

1. A. Wilbraham & J.C. Drummond, *The Englishman's Food*, 1957, p. 214; [Dr William King], *The Art of Cookery In Imitation of Horace's Art of Poetry*, 1708, pp. 3–4; Hannah Glasse, *The Art of Cookery Made Plain and Easy*, 1751 ed., Preface; E. Smith, *The Complete Housewife*, 1753, Preface; Stephen Mennell, *All Manners of Food*, 1987, pp. 83, 88–9, 125-7.
2. John Worlidge, *Systema Horti-Culturae*, 1675, p. 175; G. Scott-Thomson, *Life in a Noble Household*, 1937, p. 150; Miles Hadfield, *A History of British Gardening*, 1985, p. 127.
3. John Laurence, *The Clergy-Mans Recreation*, 1717, Preface; George London & Henry Wise, *The Complete Gard'ner*, 1719, p. i; George W. Johnson, *A History of English Gardening*, 1829, p. 147; Lorna Weatherill, *Consumer Behaviour & Material Culture 1660–1760*, 1988, p. 103.

annual expenditure on gardens by the middle classes at this time. The gardens of the gentry were not all for ostentatious display—most included a kitchen garden where fresh vegetables and wall-fruit were grown. Stephen Switzer, an early eighteenth-century seedsman and nursery gardener, observed, 'whoever has a good Cook, or reads the Books which are published in that Art, will soon find...how much...Plants that grow in a Garden contribute to the making a good Dinner; how much, if moderately us'd, to Life and Pleasure itself.' Many of the gentry migrated to London for the social Season in the winter months and, although some produce was no doubt sent from gardens at home, most of their accustomed vegetables were bought from market gardeners (Bradley, quoted above, refers to bills for 'one Winter's standing'). This demand stimulated high-quality production.[1]

Part of the motivating force pushing vegetable consumption forward was the publishing trade; in books and periodicals, people of some wealth read what they should be wearing, sitting on, hanging on the wall, planting in the garden, and, from the pages of more and more books on cookery and household management, what they should eat and how it must be prepared and presented. From an examination of a selection of the more than one hundred published cookery books of the late seventeenth and early eighteenth centuries, some conclusions can be drawn on the consumption and taste for vegetables in the middle and upper classes.[2]

Cookery books are an indirect indication of diet, the authors' choice of recipes not necessarily being what readers in fact ate. Book purchasers, however, would not buy what they did not intend to use. The books were aimed at a narrow section of society, the middle and upper classes, but were often, within those bands, passing on the tastes of those at the top of the scale to those slightly below. Thus Lamb's *Royal Cookery* outlined the dining habits of the royal house-

1. Stephen Switzer, *Country Gentleman's Companion*, 1732, pp. v-vi.
2. Virginia Maclean, *A short-title catalogue of household and cookery books, 1700–1800*, 1981.

hold and leading aristocracy to an upper-middle-class readership, whilst Hannah Glasse's clear and logical text was for the lower middle classes, as was Sarah Harrison's *Housekeeper's Pocket Book*, which contained fashionable dishes whose ingredients, 'might be purchased at a moderate Expence'.[1]

The books have some limitations for the historian of vegetables. Many advise how to choose meat and fish at markets; few comment on vegetables. More seriously, many mention vegetables rarely, particularly those little influenced by French cooking, while others discuss them in some variety. Maybe the basic English way of cooking and presenting vegetables was so simple and straightforward as to merit no mention in all but the most methodical texts. Where general directions are given, they are very simple.[2]

The Compleat Houswife (1753) offers the following advice on 'how to dress Greens, Roots, etc.':

> When you have nicely picked and washed your greens, lay them in a cullender to drain, for if any cold water hang on to them they will be tough; then boil them alone in a copper sauce-pan, with a large quantity of water, for if any meat be boiled with them it will discolour them. Be sure not to put them in till the water boils.

Individual directions are then given for preparing ten vegetables; for French beans the advice is:

> Take your beans, string them, cut them into two, and then across, or else into four, and then across, put them into water with some salt; set your sauce-pan full of water over the fire, cover them close, and when it boils put in your beans, with a little salt. They will be soon done, which you may know by their being tender; then take them up before they lose their fine green, and having put them in a plate, send them to table with butter in a cup.[3]

---

1. Patrick Lamb, *Royal Cookery*, 1710; Glasse, op. cit.; Sarah Harrison, *Housekeeper's Pocket Book*, 1743, p. iv.
2. Glasse, op. cit., pp. 15-18; E. Smith, op. cit., pp. 14–16.
3. E. Smith, op. cit., pp. 14–16.

Detailed instructions on making salads are rarely found in the cookery books, but commercial gardeners were producing lettuces, cucumbers, cress, celery, radishes, chervil, and other fleshy herbs, in abundance. Stephen Switzer wrote in 1727, 'It would be endless for me to enumerate the improvements that have been made in lettuce, and all the salletings'. John Evelyn's *Acetaria* of 1699 stands out as the only work to concentrate on salad making. Was Evelyn so comprehensive (and long-winded) on the subject that no-one wished to add to his work? Or were salads, like boiled vegetables, considered too obvious an item of food to spend time on? In *Adam's Luxury and Eve's Cookery* (1744), a book devoted solely to the growing and cooking of vegetables, 'To dress a Salled' takes up less than half a page at the very end of a chapter entitled, 'Receipts which could not come under any of the foregoing Heads'.[1]

Some widely available vegetables were scarcely mentioned in cookery books of the period. There were few or no references to roots (carrots, turnips, parsnips, and beets).[2] Books where they were included were written for the lower middle classes or country people. *Adam's Luxury and Eve's Cookery*, with many recipes for roots, advertised its contents as 'a great Variety of cheap, healthful, and palatable Dishes...Designed for the Use of all who would live Cheap, ...particularly for Farmers and Tradesmen in the Country.' Sarah Harrison's book provided a bill of fare full of root dishes for 'the frugal Mistress of a family' with ingredients which 'might be purchased at a moderate expense'. Roots were the cheap food of the London poor and were shunned by many who could afford better. A rare mention of them in the distinctly aristocratic *Royal Cookery* of Patrick Lamb[3] is apologetic: 'Beets, Are a Sort of Root, that for

1. Stephen Switzer, *Practical Kitchen Gardener*, 1727, p. vii; *Adam's Luxury and Eve's Cookery*, 1744 (reprinted 1986), p. 187.
2. William Salmon, *The Family-Dictionary*, 1705; Lamb, op. cit.; Nathaniel Bailey, *Dictionarium Domesticum*, 1736; *The Pastry Cooks Vade Mecum*, 1705; *The Compleat Cook and A Queens Delight*, 1671 ed.; *Complete Caterer*, 1701.
3. Lamb, op. cit.

being common ought not to be despised'. As early as 1629, memories of the famine led John Parkinson to remark that although Turnips are 'often seene as a dish at good mens tables,...the greater quantitie of them are spent at poore mens feasts.' A miserly gentleman in a comedy of 1632 was despised for feeding his family on roots and livers.[1]

Roots, cabbages and beans came to London markets by the cartload in enormous quantities after the Civil War; at between 2d and 6d a cartload, the toll from turnip carts at Leadenhall Market alone was £260 in 1696, and in the middle of the eighteenth century roots were still a staple of the poor. The lowly status of the root was reinforced by its increasing use as animal fodder; turnips were, by the middle of the eighteenth century, as likely to be eaten by cattle and sheep as people.[2]

Despite their inferior status, roots were eaten in some quantity by the middle classes, as Misson and Kalm recorded. But they were not the vegetables favoured by those who could afford better, and they were definitely not fashionable.

What vegetables were most prized? Those which occur frequently in the cookery books examined are: artichokes, asparagus, cauliflowers, cucumbers, French and kidney beans, green peas, lettuce, mushrooms, and spinach, with broccoli gaining in popularity later in the period. These may be considered the 'superior' vegetables of polite London society, and the recipes in which they were included may explain their appeal.

Some vegetables—asparagus, artichokes, and lettuces in particular—had been firm favourites for a long time. Throughout the

1. John Parkinson, *Paradisi in sole*, 1629, p. 509; J. Massinger, *The City Madam*, 1632, Act I Scene i.
2. P.V. McGrath, 'The Marketing of Food, Fodder, and Livestock in the London Area in the Seventeenth Century', (unpublished thesis, Univ. of London, 1948) pp. 195–6; H. Le Rougetel, *The Chelsea Gardener*, 1990, p. 137; Malcolm Thick, 'Market Gardening in England and Wales' in *The Agrarian History of England and Wales*, vol. V part 2, 1985, edited by Joan Thirsk, p. 532.

seventeenth century they occur in cookery books and other literary sources. Ben Jonson mentions artichokes and lettuce in 1616. Both a fictional prostitute in the 1630s and Samuel Pepys in the 1660s enjoyed asparagus. Maybe such vegetables were liked simply because their texture and taste found widespread satisfaction; in eighteenth-century terms, they were 'delicate'. (This widely used epithet also connoted superiority and gentility.) These, and most of the other superior vegetables, could be served in many ways or had particular virtues which were widely exploited in cooking. Asparagus was boiled and buttered, covered in various sauces as a side-dish, extensively used to flavour stews, used as a pie ingredient and as a garnish for chicken and other dishes. Lettuce was the basis of many salads and was a frequent ingredient in cooked dishes. Cauliflower was a favourite garnish. Mushrooms flavoured and coloured ingredients and were a mainstay of French cookery; the Muse in the poem *The Art of Cookery* was instructed to sing of 'the man that did to Paris go, That he might taste their Soups, and Mushrooms know.'[1]

Broccoli, increasing in favour in the early eighteenth century, was relatively new to England, although mentioned as a delicate vegetable by Evelyn in 1699. The seed was imported from Italy. Bradley wrote in 1723: 'tis a plant which has been cultivated privately in some few Gardens in England, for about three years', but commercial production had not begun. He planned to produced a pamphlet on its culture. Stephen Switzer gave pride of place to it in a small work on new vegetables in 1728. Broccoli is a good example of the eighteenth-century quest for novelty which led both to the introduction of new vegetables to tempt the rich and to the development of new varieties of established crops. Some new varieties had little

1. Ben Jonson, *Epigrammes*, 133, 1616; John Murrell, *Two Books of Cookerie and Carving*, 1638; *The Compleat Cook and A Queens Delight*; *The Court and Kitchen of Elizabeth, commonly called Joan Cromwell*, 1664; Massinger, op. cit., Act III Scene i; Charles Cooper, *The English Table in History and Literature*, 1929, p. 65; [William King], *The Art of Cookery In Imitation of Horace's Art of Poetry*, 1708, p. 79; *The Diary of Samuel Pepys*, ed. Wheatley, 1946, vol. VI, p. 263.

to commend them but novelty. Bradley complained: 'We have great varieties of Lettuce, many of which have been more esteemed for their Rarity then for their Goodness.' Evelyn, in 1699, could remember when melons were rarely cultivated and 'an ordinary melon would have been sold for five or six shillings'; in 1744, however, it was suggested: 'to enumerate all the different Sorts of this Fruit, would be not only endless, but impossible, there being annually new Sorts brought from abroad, a great many of which proved good for little.'[1]

A final characteristic of 'superior' vegetables is that they were, for much or all of the year, more expensive than roots. Richard Bradley's market reports bear this out and it is seen clearly in J.C. Loudon's survey of Covent Garden prices in 1822, which reveals that although the price of roots such as carrots varied little over the year, rising only with the new crop, some other vegetables were unobtainable for parts of the year and, at other times, commanded a high price. This premium was occasioned by the desire of the rich for vegetables when they were scarce, a distinctly modern-sounding fashion which was nonetheless well known over a century before Loudon's statistics.[2]

In 1684, demand for out-of-season vegetables attracted moral censure: 'And verily the vanity of some deserves our wonder, who are of that Heliogabalian Stomach, to which nothing doth relish which is not dear...onely loving Pease, when they are scarce to be had.' Evelyn was contemptuous of forced 'Early Asparagus...so impatiently longed after,' and Steele, in *The Tatler*, in 1710 summed

1. John Evelyn, *Acetaria*, 1699, pp. 16, 38; Bradley, *A General Treatise of Husbandry and Gardening*, 1726, p. 43; Bradley, *New Improvements of Planting and Gardening*, 1719, p. 164.; Stephen Switzer, *A Compendious method for raising Italian broccoli*, 1728; *Adam's Luxury and Eve's Cookery*, p. 48.
2. J.C. Loudon, *Encyclopaedia of Gardening*, 1824, p. 1062. Loudon's statistics are outside our period but, although the population of London was then much larger and some tastes had changed, the basic conditions of supply and demand remained much the same as in the early eighteenth century. Bradley, *A General Treatise,* pp. 41–4, 108.

up the fashion by observing: 'They are to eat every Thing before it comes in Season, and to leave it off as soon as it is good to be eaten.' In 1719 Richard Bradley could say: 'the Pride of the Gardeners about London chiefly consists in the production of Melons and Cucumbers at times either before or after the natural Season.'[1]

Explaining his publication of the fancy prices charged for unseasonable vegetables, Bradley wrote, 'I choose to mention the Prices of these Curiosities, that we may the better judge of their Scarcity, and compare them with Fruits of the same kind another Season.' In May 1723 he observed, 'Forward Pease were sold this Month for Half a Guinea per Pottle-basket' (equivalent to seven guineas a half-sieve, compared with 6d for the same measure in July 1822). Loudon's survey shows a steep decline in the price of peas as the main crop succeeds those first harvested, and the pea suffered a similar annual drop in culinary status, sold in quarts to the rich in May or June, and shovelled off the tail of carts to poor customers in high summer. In June 1723 Bradley reports: 'About the Middle of the Month, most of the Crops of Pease and Beans about London were ripe, and came daily in such Quantities to the Markets, that their Price was reduced to about one shilling per Bushel.' In May also, the rich could buy 'Collyflowers, of the right sort ... for 5s. each'. He reported that 'Kidney-Beans raised in Hot-Beds were about 3s. or 4s. per Hundred' (in March, raised in the same manner, they had been 2s. 6d. a dozen), and 'Mushrooms were bought for eight and ten Shillings a Basket, in St. James's Market' on March 27 1723.[2]

Stephen Switzer summarized the progress made in artificially lengthening the growing seasons; progress largely achieved by commercial gardeners:

1. *The Compleat Tradesman*, 1684, p. 17; Evelyn, op. cit., pp. 134–5; *Tatler*, March 21st, 1710; Bradley, *New Improvements*, *ut sup.*, p. 116.
2. Bradley, *A General Treatise*, *ut sup.*, pp. 41–4, 108, 148; Loudon, ibid.; 'Peas Sold from a Cart', wash drawing by Robert Dighton, *c.*1786, Museum of London.

Who then, till within these few years, could have imagin'd that the cucumber, which seldom was seen heretofore (even since my remembrance, who have not been above twenty five years a practitioner in Gardening) Till the middle, or perhaps the latter end of May, seldom the beginning, that are now produced in and about London, and several places in the country, in the beginning of March; and the industrious among the Gardeners are still striving to outvie one another, and will in all probability produce them in February, or sooner, and that as good or better than they have in any of the succeeding months, when they have less time to tend them.

And as the fruits that grow in the kitchen garden are so much more accelerated now than they were heretofore, so are the legumes and herbacious rooted plants, the collyflower in particular, that never shew'd its beautiful head above three of four months in the year, appears now above six or seven furnishing the tables of the curious all that while with its wholesome nourishment; and by good management mocks the severity of our unsteady climate.

The phaseolus, or kidney bean, that used not (but was thought too tender) to be sown till the beginning or middle of April, is now, by the means of frames and glasses, and that with little trouble, sown in January and February; and the fruit (if it may be so called) which used to be fit to gather heretofore not till the middle of June, is now fit for the table by the beginning of April; and which is more, by the great skill and improvement of our industrious Gardiners it continues a constant and most useful dish for every week in the year between that and the beginning of October.

Even peas and beans, that were heretofore the produce but of two or three months, furnish the table with an agreeable dish for seven or eight; viz. from April to almost Christmas; so expert are our Gardiners now in the retardation of the produce of the Garden, as well as in the bringing of it in early.[1]

1. Switzer, *The Practical Kitchen Gardener*, 1727, p. vii.

Around London market gardeners were producing fashionable vegetables for genteel tables while others were growing bulky roots for general consumption. Social interaction was important in attuning gardeners to the market as well as a determinant of taste. These fashions travelled down the social scale, but a more subtle influence on eating habits in London came from below, from the gardeners themselves, as they produced new varieties to tempt the palate or produced familiar vegetables in unfamiliar seasons. London was at the centre of this desire for novelty and change. Each winter the leading gentry went to the capital for the season, a round of social gatherings where changes in fashion spread very quickly. Bernard Mandeville, summing up the influence of fashion included both gardens and food in a list of desirable goods:

> The worldly minded, voluptuous and ambitious Man, notwithstanding he is void of merit, covets Precedence every where, and desires to be dignify'd above his Betters: He aims at spacious palaces and delicious Gardens...His Table he desires may be served with many Courses, and each of them contain a choice variety of Dainties not easily purchas'd, and ample evidences of elaborate and judicious Cookery.[1]

1. Bernard Mandeville, *The Fable of the Bees*, 1717 (ed. Phillip Harth, 1970), pp. 170–1.

CHAPTER II

The spread of market gardening around London

'In some seasons the gardens feed more people than the field...'

London was by far the largest city in England in 1550 and its population increased rapidly over the next quarter-millennium. Only in the nineteenth century did industrialization create other sizeable English conurbations. Estimates of population show the pace of growth:

1500	50,000	*1700*	575,000
1550	70,000	*1750*	675,000
1600	200,000	*1880*	900,000[1]

This concentration of people had to eat—to buy each day necessities or luxuries, according to individual pockets. Rapid growth placed huge demands for food on the surrounding countryside. By 1700, farmers sent produce to London from many parts of England and a ring of market gardens around the capital spread deeper into the surrounding counties as demand grew, as buildings displaced gardens sited closer to the city, and transport improved. So effective was the provision of garden produce in response to this demand that by the 1660s, one observer found it 'incredible how many poor people in London live thereon, so that in some seasons the gardens feed more people than the field.'[2]

1. Roy Porter, *London, A social history*, 1994, pp. 42, 97–8.
2. Malcolm Thick, 'Root crops and the feeding of London's poor', in *English Rural Society*, ed. J. Chartres & D. Hey, Cambridge, 1990, p. 295; Thomas Fuller, *The Worthies of England*, 1662, p. 7.

Figure 2. Possible market gardens near the manor of Paris Garden across the Thames from the City of London, shown in the Agas map of 1562.

As befits a major city, London has, from the mid-sixteenth century, been mapped with increasing regularity. With some interpretation, as none before Thomas Milne's land use map of 1801 specifically identifies all market garden land around the capital, these surveys provide a quick and informative overview of horticulture at the moment of their production.[1]

The earliest complete map of London is that of Braun and Hogenberg, a small, one-sheet engraving first published in a German atlas in 1572, showing the City, Westminster and their immediate environs. It is probably a copy of a map engraved in copper in the 1550s of which only part survives. It is thought reasonably accurate and well drawn. It shows areas of market gardening at Houndsditch near the Tower of London; Spitalfields; beyond Moorgate; to the north of Gray's Inn in Holborn; and south of the Thames in Southwark near Paris Garden Stairs.

Ralph Agas's map of about 1562 (figure 2), although on a much larger scale, is probably a copy of the same lost copper-plate map. It shows similar areas of market gardening: 'patches' to the west, north, and east of the City each made up of several gardens, a pattern which becomes increasingly common in the two succeeding centuries. Only south of the river was gardening still little in evidence.[2]

Little full-time market gardening in London has been found before 1500, although there had been a regular market for garden produce near St Paul's in 1345 where any surplus from noble and monastic gardens was sold. Contemporary observers believed market gardening began to increase from the start of Henry VIII's reign. The historian John Stow recalled from his youth a gardener

1. I discuss this problem in relation to the Neat Houses, below, and believe that it is possible to discern from maps which land was gardened commercially. Some mapmakers used particular symbols for commercial garden ground. Others depicted all gardens similarly but the size of the garden, the type of housing on it, its relationship with known commercial gardens, backed up by documentary evidence of commercial gardening at the time, allow reasonable assumptions about what is a market garden on the maps to be made.

2. *A Collection of Early Maps of London, 1553–1667*, intr. by John Fisher, 1981.

Figure 3. A market garden beside the Lambeth Road near Lambeth Palace in the 1660s. The garden is shown surrounded by a paling fence. From Wenceslaus Hollar, *The Prospect of London and Westminster taken from Lambeth.*

named Cawsway who, sometime before 1552, had a garden at Houndsditch near the Tower and 'served the Markets with Herbes and Rootes.' He remembered another early commercial garden nearby in the Minories. Stow also makes reference to what could well have been commercial gardens operating in the 1590s at various other places in the suburbs: near Tower Hill, off Goswell Street north of Charterhouse, and at Shoreditch. His observations tie in with the maps.

Documentary evidence has been found of market gardening in Edmonton Hundred, north of London, in the second half of the sixteenth century and the first Dutch and French refugees were running gardens in various Southwark parishes by the 1570s. In 1597 the herbalist John Gerard mentions Hackney as a village where the best turnips were grown for sale in the City.[1]

The development of market gardening around London in the first half of the seventeenth century consequent on the impetus provided by the emergency of the famine years was reflected in the incorporation of the Gardeners' Company in 1605. The Dutch gardeners who introduced gardening to the Surrey side of the Thames had stimulated a sizeable industry there by the middle of the century.

1. H.T. Riley, *Memorials of London Life*, 1868, pp. 228–9; John Stow, *A Survey of London*, ed. Henry Morley, n.d., pp. 15, 149, 385; '"Now turned into fair garden plots" (Stow)', J.G.L. Burnby & A.E. Robinson, Edmonton Hundred Historical Society Occasional Paper, n.s. 45, pp. 4–5; John Gerard, *The Herbal, or General Historie of Plantes*, 1633, p. 232.

Commercial gardeners also flourished to the east of the City. The records of Middlesex Quarter Sessions contain the names of eight gardeners from Whitechapel, five from East Smithfield, and four from Stepney between 1612 and 1666.

Parallel changes were occurring west of the City. The first gardens at the Neat Houses in Westminster started around 1600. Fulham was renowned for parsnips by 1610. Market gardening was inserted into the open field system there with such success that the Gardeners' Company deemed the husbandmen 'gardeners' and sought to bring them within its jurisdiction (which extended for 6 miles around the City). They began legal proceedings in 1616 and extended the action to farmers and gardeners in Chelsea and Kensington by the time the case was finally settled in 1633.[1]

By 1655, London had 'Gardening-ware in...plenty and cheapness', an impression confirmed by Faithorne and Newcourt's map published in 1658, the first new survey that century. Crudely drawn, it is nevertheless accurate and large-scale, although the area covered is only slightly larger than the maps from the previous century. At first sight London seems ringed with orchards but the rows of stylized trees occur in known market garden areas and also, apparently, indicated commercial gardens.[2]

If we begin our survey of this map in the north-western quarter, it plots gardens around St Giles church and others north of Holborn. More are found along the northern edge of the built-up area in Clerkenwell, north of Moorfields and in Shoreditch. Gardens to the east of the City are visible around Stepney and on both sides of Ratcliffe Highway. South of the river, a ribbon of market gardens sited in closes beyond the buildings stretches from Bermondsey almost as far as Lambeth, with a particular concentration in St George's, Southwark. Gardens are also noticeable around Horsely Down,

1. See p. 21, above, and the references there.
2. Samuel Hartlib, *His Legacie of Husbandry*, 1655, p. 9; *Early Maps of London, ut sup.*, plates 11–18.

Southwark. This was a particularly early area of commercial gardening: Gerard mentions a nurseryman there in 1597.[1]

To the west of London there were gardens on the fringes of Westminster near Tothill Fields and near Tart Hall to the north. Like all early maps of London and Westminster, Faithorne and Newcourt's does not extend to the Neat Houses, nor to the villages further into Middlesex where gardening was expanding at this time, but the area of probable market gardening is much greater than that recorded by the maps drawn in the mid-sixteenth century.[2]

A further measure of the extent of market gardening in London by the time of Faithorne and Newcourt can be taken from an analysis of the quarterage payments to the Gardeners' Company for 1661–2 and 1688–9, recording membership within its jurisdiction. Of 202 identifiable addresses in 50 different parishes in 1661–2, 108 were south of the Thames, 47 west of London, 37 to the east, and 10 to the north. The 1688–9 quarterage records list fewer gardeners and in only 26 different parishes. When, however, places with five or more recorded gardeners in 1661–2 are compared with the same locations in 1688–9, a clear pattern of concentration emerges. Battersea and the Neat Houses were the most important areas in both years, but there were also significant numbers at Lambeth, Greenwich, Deptford, Southwark, and Putney south of the river; Shoreditch, Whitechapel and Ratcliffe Highway to the east; and at Westminster and Fulham to the west. Only in the North were there few gardeners in the Company, although the odd member worked at such places as Moorfields, St Giles, or Islington.[3]

Ogilby and Morgan's map of the City in 1676 (figure 4), drawn to a large scale from an actual survey, does not cover much of the London suburbs. It extended south only to the Thames, to Lincoln's

1. John Gerard, *The Herbal, or General Historie of Plantes*, 1633.
2. John Harvey, *Early Nurserymen*, 1974, p. 41; Martha Carlin, 'Four plans of Southwark', *London Topographical Record*, XXVI., p. 21; Moses Glover, *Map of Isleworth and Twickenham*, 1635.
3. Guildhall Library, MS 21,129/1; MS 21,129/8.

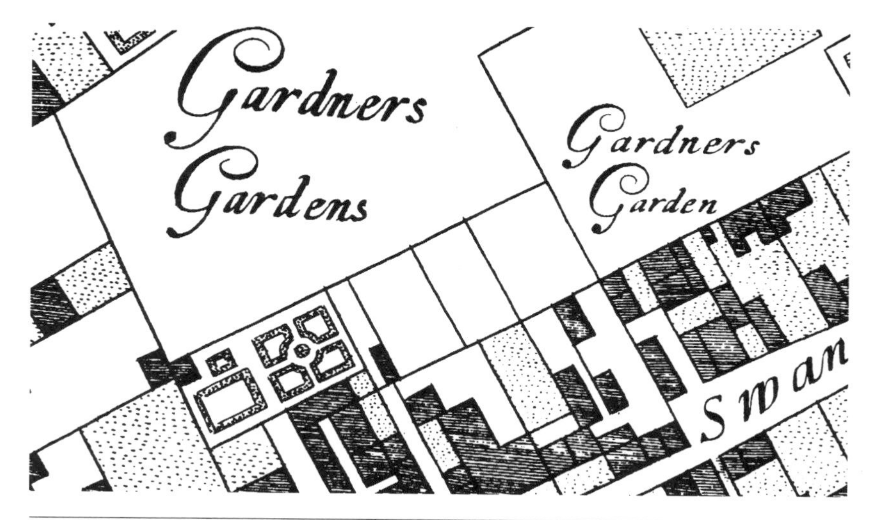

Figure 4. Market gardens ('Gardners Gardens') in the suburb of Clerkenwell, north of the City. From Ogilby and Morgan.

Inn in the west, the Tower in the east and Clerkenwell and Shoreditch in the north. Bricks and mortar covered almost all of this area. Open ground accommodated market gardening just north of Grays Inn, in Clerkenwell, and in Shoreditch. These mapmakers unambiguously labelled market gardens 'Gardners Garden' and the map confirms the sites of gardens tentatively identified earlier.[1]

Joel Gascoyne's survey of Bethnal Green in 1703 (figures 5 & 6) confirms information gleaned from other maps. Gascoyne depicted the extensive market gardening he found in a conventional manner as rows of beds filled symbolic vegetables. He drew a distinction between private and commercial gardens by naming the commercial gardeners.[2]

Commercial gardening expanded fast after the Restoration. John Aubrey wrote, 'I may afirm that there is now, 1691, ten times as

1. *The A to Z of Restoration London*, intr. Ralph Hyde, London Topographical Society Publication 145, 1992.

2. *An Actual Survey of Bethnal Green, Stepney, etc.*, Joel Gascoyne, 1703, Bodleian Library, Oxford, Gough Maps London, 18.

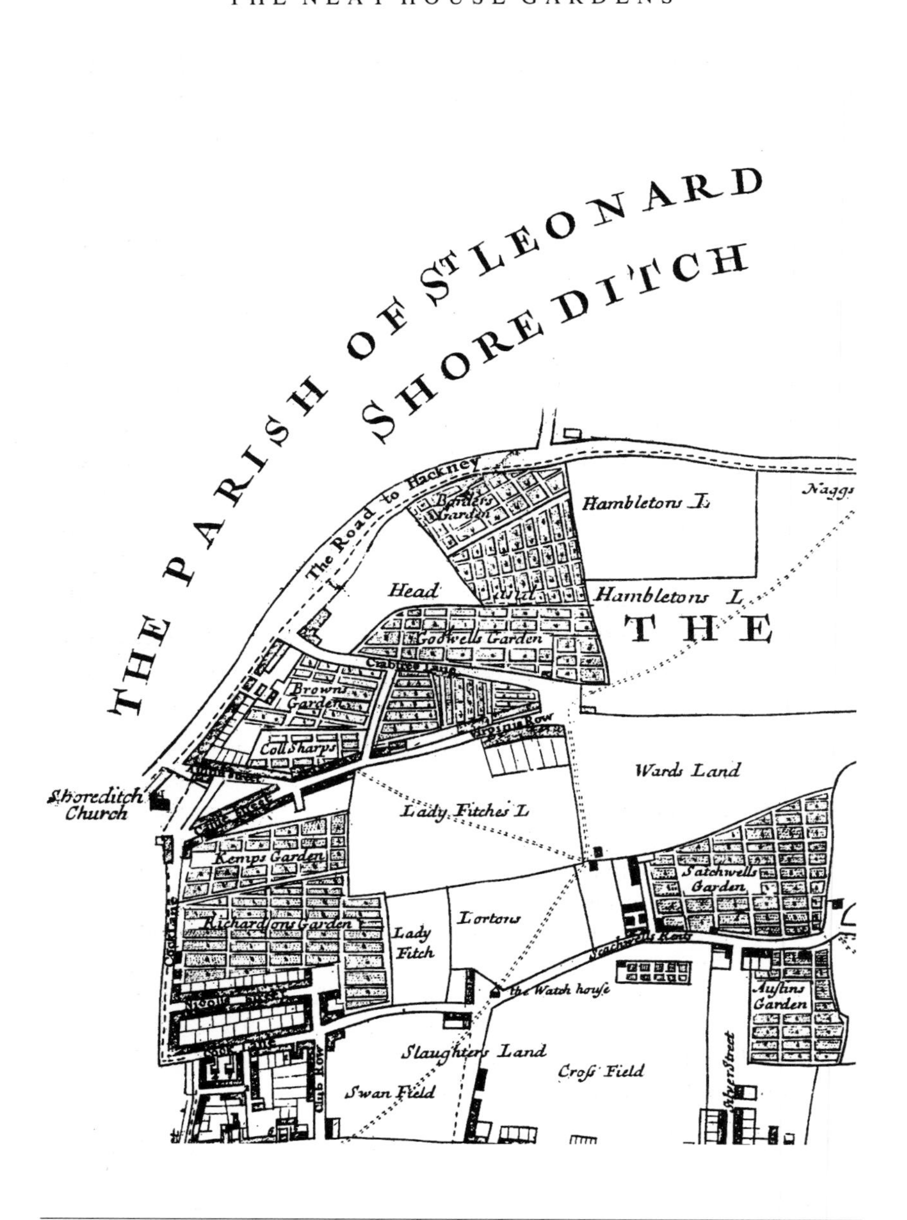

Figure 5. Market gardens near Shoreditch, 1703, from Joel Gascoyne's survey.

much gardening about London as there was in 1660.' Richard Bradley considered an eleven-fold increase in acreage had taken place between 1688 and 1721, by which time he estimated there to be 100,000 acres of market gardening. Whilst such broad estimates should be treated with care, the trend is clear.

This period saw the golden age of the Neat House gardens, when they became renowned for advanced techniques. Significant developments also occurred as husbandmen in Fulham, Chelsea, and Kensington perfected their mixture of farming and gardening. Although some grew more delicate produce, such as asparagus, in the open fields, they concentrated on traditional field crops and roots for the poor. In 1704, one Fulham farmer-gardener with a holding of 25 acres of open field was accustomed to grow nine acres of peas, three acres of wheat, nine acres of barley, three acres of carrots, and sometimes half an acre of tares. Careful cultivation and manuring for vegetables produced good ground for succeeding grain crops but the profitability of vegetables was such that more land was taken up with gardening in these parishes as the century progressed. In 1810, the agriculture in Chelsea Fields was described as follows:

> The most perfect, and best cultivated culinary grounds are in this parish and its vicinity; and here, in general, the characters of farmer and gardener are united in the same person, as the grounds are successively filled with grain and vegetables.
>
> In the months of January and February they crop with early pease, to be gathered in the month of June. In a few days afterwards the ground is cleared, the pease haulm stacked up for future fodder and the plough being set to work, the land is sown with turnips, which are sold off again in autumn, when the ground is again ploughed, and filled with colewarts for the spring use....Every gardener has a favourite and particular system in the succession of crops. [1]

1. Miles Hadfield, *A History of British Gardening*, 1960, p. 127; Richard Bradley, *A General Treatise of Husbandry and Gardening*, 1723, vol. II, p. 184; GLRO, MI 1678/11; MI 1702/30; PRO, E134/2&3 One Hil.1; Thomas Faulkner, *Historical and Topographical Description of Chelsea and Environs*, 1810, p. 14.

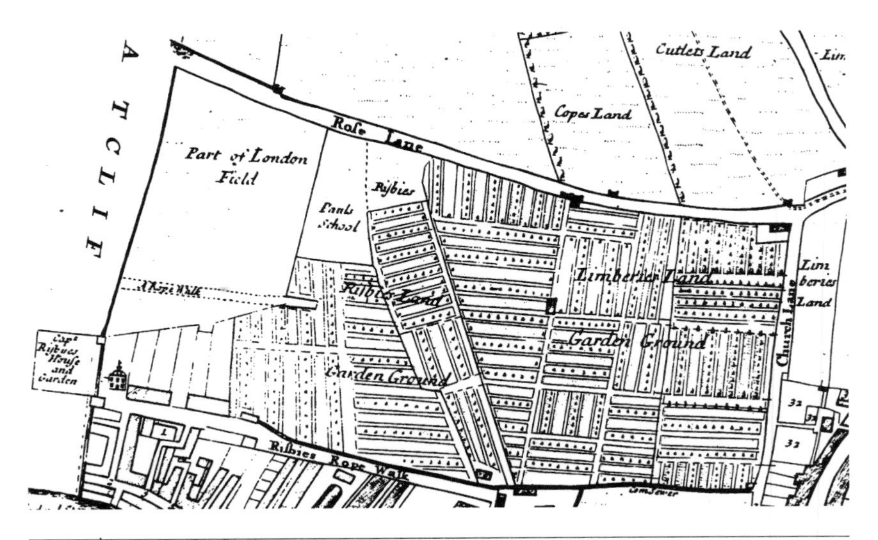

Figure 6. Market gardens near Limehouse, 1703, from Joel Gascoyne's survey.

On the Surrey bank of the Thames at Barnes, Putney, Newington, Richmond and Lambeth farmers copied neighbouring kitchen gardeners and incorporated an acre or two of parsnips, carrots, turnips, garden peas and beans or maincrop asparagus into their rotations. The demand from London spread this farm gardening deeper into Surrey: in the 1680s the vicar of Croydon took court action to secure the tithes of root vegetables introduced into the open fields there. By the middle of the eighteenth century, the specialist garden crop liquorice was grown in the deep Croydon soils.[1]

Tithe disputes, a sure sign of changes in agriculture, also occurred some miles to the west of London, as nursery and market gardening moved up the Thames and deeper into Middlesex. In the course of a dispute at Twickenham and Isleworth in the 1660s, one tree nurseryman said that he sold 2,000 trees in 1662. His son, Peter Mason

1. Lambeth Palace Library, UH 90/688; UH 90/2945; UH 96/2730; UH 96/414; UH 96/1062 ; PRO, PROB 5/2138; PROB 4/3178; E 134/33&34, Chas II, Hill.26; *Museum Rusticum*, 1764, vol. II, p. 256.

the younger was said in 1728 to have 'one of the best Collections of English Fruits of any Nursery-Man in England'. On his death two years later, his nursery contained over 100,000 trees both common and rare. In 1724, the vicar described the growth of gardening in this parish:

> The Town and parish of Isleworth is situated near the River of Thames and by reason of this easy and convenient water carriage from thence to London and other places from that River great parts of the Common fields and Closes of Isleworth have been Inclosed and Converted into Gardens orchards and nurseries of Fruit trees and Greens and that within 50 or 60 years since past several hundred acres of arable or Common Field Grounds of the said parish have been Converted into such gardens & Inclosures.[1]

During the eighteenth century the size of London's market gardening hinterland continued to expand. The produce and methods of cultivation became ever more diverse. What follows is a bare outline of developments.

The ingenuity in the face of a growing demand for novelty from rich customers is captured by this description of artichoke production by Stephen Switzer in 1727:

> It is a plant that is cultivated amongst market-gardiners about London, with more than ordinary industry, because it brings in great profit, for about and after Michaelmas all their whole gardens at Rotherhith, Lambeth, and other adjacent places, are nothing else; where putting them into a kind of basket they call maunds, they sell them from two, to three, four, or five shillings per maund, that does not hold above a dozen at most, fewer or more according as the artichokes are in size; those that are the largest being the most valuable, as yielding what they call the largest bottoms, and consequently the most meat.[2]

1. PRO, E.134, 16 Chas II, East 1; E.126/99, 2 May 1664; E.126/23 Mich 1724.
2. Stephen Switzer, *Practical Kitchen Gardener*, 1727, p. 154.

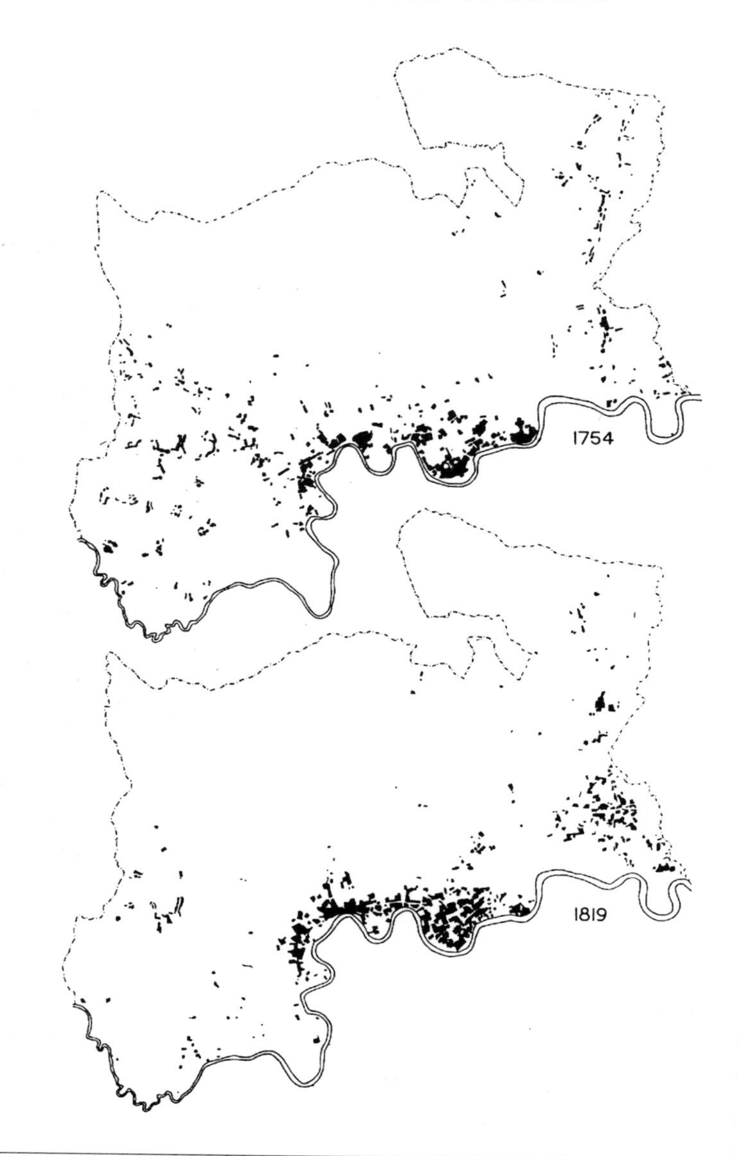

Figure 7. Market gardening in Middlesex in 1754 (after Rocque) and 1819 (after Greenwood). From *Middlesex and the London Region* (part 79 of *The Land Utilization Survey of Britain*), E.C. Willats, 1937.

A few years before this was written, Richard Bradley listed areas where some garden delicacies were grown:

> Battersea affords the largest natural Asparagus, and the earliest Cabbages. Again, the Gardens about Hammersmith are as famous for Strawberries, Raspberries, Currants, Gooseberries, and such like; and if early Fruit is our Desire, Mr Millet's at North End, [Fulham] near the same Place, affords us Cherries, Apricocks, and Curiosities of those kinds, some Months before the Natural Season.[1]

Rocque's map of London, Westminster and Southwark, published in 1746 (the same year as his map of London and ten miles around), was the first based on a new survey for over fifty years. This large-scale map covers the built-up area and suburbs as far as Marylebone and Chelsea in the west, Clerkenwell and Shoreditch in the north, Stepney and Limehouse to the east and a broad band south of the river from Lambeth to Deptford. The map includes, in great detail, large areas of market gardens on all sides of London. They are scattered along the northern edge of the built-up areas, and blocks of gardens can be seen in the east, clustered north of Old Street, Hoxton, Spitalfields, Whitechapel, along Mile End Road and at Bethnal Green. Stepney was surrounded by gardens. And they were also to be found at Shadwell, Limehouse and along Ratcliffe Highway. In the south-west, gardens encroached on both sides of Tothill Fields in Westminster and the mass of gardens at the Neat Houses are clearly defined, together with those on the edge of Chelsea (figures 7– 10).

The largest acreage of gardens on the map was south of the Thames: beyond the fringe of housing by the river, the open fields and closes of Lambeth, Southwark, Kennington, Newington, Walworth, Bermondsey and Deptford were considerably given over to this new form of agriculture.[2]

---

1. Richard Bradley, *Philosophical Account of the Works of Art and Nature*, 1721, p. 184.
2. *The A to Z of Georgian London*, ed. Harry Margary, 1981.

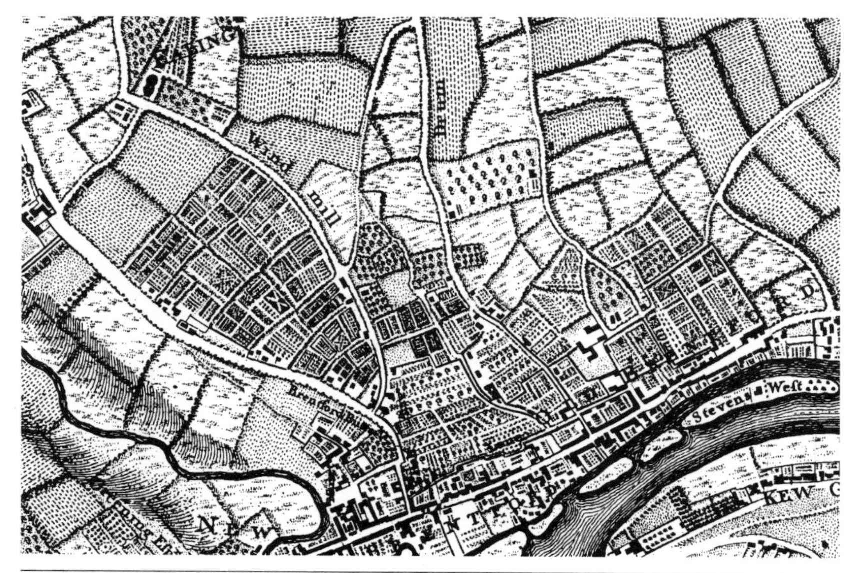

Figure 8. Market gardens spreading into the fields north of Brentford, west of London, in the 1740s. From John Rocque, *An Exact Survey of the Cities of London and Westminster*, 1741–6.

Downstream, well beyond the south-eastern edge of Rocque's map, the farmers near Gravesend in Kent also felt the pull of the London market. A commentator remarked in 1749 that:

> Within a few Years past, very great Improvements have been made in the Lands near this Town, by turning them into Kitchen-Gardens, the Land being fresh for this Purpose, also pretty moist, and the Town having good Quantities of Dung made in it, with which they Manure the Land. It produces very good Garden-Stuff in great Plenty, wherewith they not only supply the Towns for several Miles round, but also send great Parcels to the London Markets; particularly Asparagus, which is so much esteemed that the Gravesend Asparagus bears a better Price than any other, even that of Battersea.[1]

1. Thomas Read, *A New Description of Gloucestershire etc.*, 1749, p. 445.

The same attraction was felt throughout the suburban hinterland. Analysis of Rocque's map of Middlesex of 1754 shows a long tail of gardening westwards from Westminster, both close to the Thames (for ease of transport) as far as Twickenham and in scattered locations inland as far as the county border. Further extensions can be observed snaking from Limehouse near the Thames to Tottenham, Edmonton and Enfield. The land use map in the front of Middleton's *View of the Agriculture of Middlesex* of 1798 confirms this pattern.[1]

Daniel Lysons, in his survey of the parishes within 12 miles of London published in 1792, commented on the diversity of gardening. He praised the quality of cabbages and asparagus raised in Battersea by intensive methods. At Lambeth, he found nurseries as well as vegetable gardens and he also noted nurseries in Hackney specialising in exotic plants. Farmer-gardeners, who rotated field crops with vegetables, operated in Fulham, Chelsea and Bethnal Green. In Mitcham, to the south, about 250 acres were occupied by 'physic gardeners' who grew such crops as peppermint, lavender, wormwood, camomile, aniseed, rhubarb and liquorice for apothecaries to turn into medicines.

Lysons concluded that within 12 miles of London, about 5,000 acres were cultivated for vegetables, 1,700 acres for fruit, and about 1,700 acres were cropped with potatoes. In addition, about 1,200 acres of vegetables were raised as cattle fodder. (John Middleton, in 1798, estimated market gardening around the capital at 10,000 acres with another 1,500 acres of nursery ground in Middlesex alone.)

Lysons' comment that the acreage of gardens at Chelsea had fallen by 167 acres between 1664 and 1792 because of 'the prodigious increase of buildings' reminds us that London was growing continuously, absorbing surrounding villages and their farm and garden ground. Poorer land was built on first. High-quality garden ground, such as that at Fulham, Chelsea, and particularly the Neat

1. E.C. Willats, *Middlesex and the London Region* (part 79 of *The Land Utilization Survey of Britain*), 1937, p. 294.

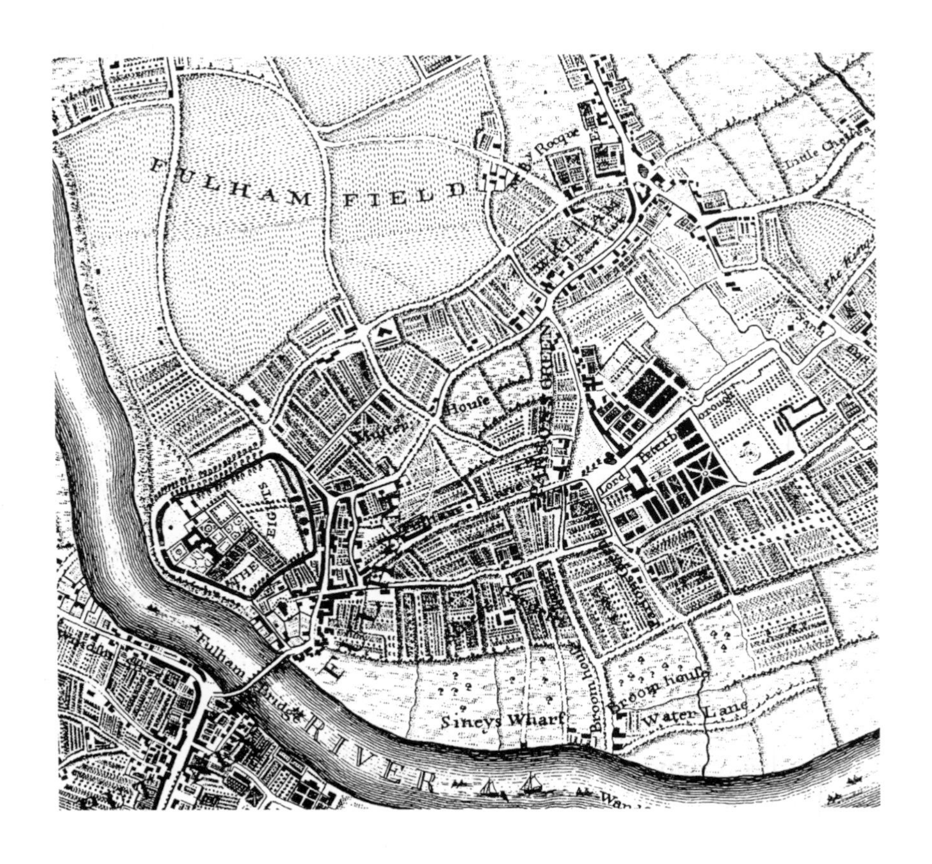

Figure 9. Enclosed market gardens at Fulham and Parsons Green, west of London, in the 1740s. Open fields ('Fulham Fields') were worked by farmer-gardeners who rotated field and garden crops. The garden labelled 'Mr By. Rocque' belonged to a nursery gardener, Bartholomew Rocque, brother of the map-maker. From John Rocque, *An Exact Survey of the Cities of London and Westminster,* 1741–6.

Houses, could command relatively high rents, thus survived when all around had been built upon.[1]

Lysons estimated land use in each of the parishes he surveyed, and from this information a table of land under market gardening in 1792 has been compiled. This total includes fields cropped with potatoes for the London markets in East and West Ham, but otherwise does not cover conventional agricultural holdings. Surprisingly, the west had still the largest acreage at this time, although the greatest concentration was in the parishes just outside the built-up area south of the river, exemplified by the high number of southern parishes with over 100 acres of garden ground. Fulham, with much of its open fields under vegetables, stands out among the western parishes.

Thomas Milne's land use map published in 1800 was probably the first to employ techniques and conventions still current today (figure 11). Fields and gardens were given key letters corresponding to land use and the map was hand painted in colours denoting the type of agriculture within each enclosure. Market gardens were blue, nurseries yellow, and whether an area was open field or enclosed was made clear. The map covered London and its environs, a total of 260 square miles reaching deep into the surrounding counties, giving a highly accurate picture of agriculture at the time. Swathes of blue stretch north of the Thames from Westminster as far as Twickenham; another band runs south of the river from Lambeth to Deptford. There is an eastern patch from Limehouse to Edmonton. The Neat House gardens stand out as a solid piece of blue, surrounded by housing.[2]

The patterns of gardening areas close to London shown on Milne's map do not differ much from those of the early seventeenth

1. D. Lysons, *The Environs of London*, 1792, vol. I, pp. 27, 68, 257, 350; vol. II, pp. 28, 71, 345; vol. VI, pp. 573–6; John Middleton, *View of the Agriculture of Middlesex*, 1798, pp. 266–7, 272.
2. *Thomas Milne's Land Use Map of London & Environs in 1800*, intr. G.B.G. Bull, London Topographical Society Publications 118 & 119, 1975–6.

Table 1. Garden acreage in parishes within ten miles of London, 1792 (D. Lysons).

West of London	*Acres*
Fulham	2175
Kensington	590
Isleworth	430
Chiswick	280
Ealing	250
Chelsea	170
Twickenham	150
Paddington	84
Acton	10
Marylebone	5
Total	*4144*
South of London	
Deptford St Paul	500
Battersea	300
Mitcham	250
Lambeth	250
Mortlake	250
Wandsworth	218
Greenwich	160
Barnes	150
Putney	120
Newington Butts	100
Plumstead	90
Bromley St Leon	60
Wanstead	50
Rotherhithe	40
Total	*2538*
North of London	
Edmonton	27
Newington Stoke	18
Islington	5
Total	*50*
East of London	
West Ham	700
East Ham	570
Barking	400
Bethnal Green	140
Little Ilford	120
Stepney	50
Hackney	30
Leyton	25
Stratford	13
Limehouse	10
Total	*2058*
GRAND TOTAL	*8790*

century: close to markets but stretching along the Thames where transport was cheap and efficient. There are reasons for this. Garden produce is often delicate and perishable. The most fresh commanded the highest prices which dictated that gardens should be close to the built-up area and the markets. Access to good transport was an advantage, and became more important the further gardens were from markets. Transport was also needed to bring back manure from the streets and privies of the city.[1]

1. This summarizes part of a theory of agricultural location put forward by the German geographer J.H. von Thünen in 1826 and generally adopted by later geographers.

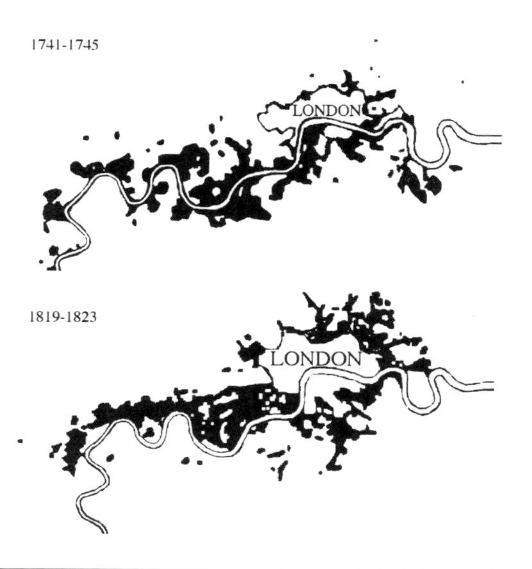

Figure 10. Market gardens near London, extracted from Rocque's survey, and county maps by C. & I. Greenwood of Middlesex, Surrey and Essex (1819-23). Land use within the London boundary has not been recorded. (From P. Atkinson, 'The Charmed Circle: von Thünen and Agriculture around 19th-century London', *Geography*, vol. 72, 1987, p. 132.)

Middleton summarized the acreage and gross sales of market gardens around London in his *View* of 1798. His statistics sit well with the information gleaned from Milne (table 2). Allowing this table to be composed of estimates, it is an impressive tally of the economic importance of production at this time. Middleton added to his figures an estimate of £400,000 as the gross income from fruit gardens near London, making a total annual product of £1,045,000.

Throughout the period, most garden seeds for the national market were sold by London suppliers and London suburbs were home to the most important nurseries of flowers, plants and trees. Early nurseries were small and had a narrow range of stock. The earliest were concentrated on Westminster and to the east of the City, but by the end of the seventeenth century, larger general nurseries had been founded, growing more types of garden plant as new species were

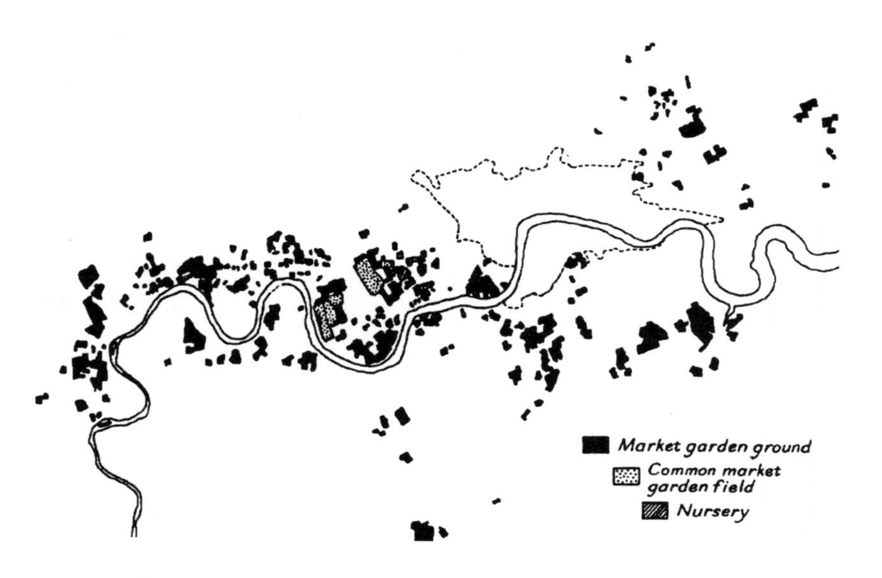

Figure 11. Market gardens around London in 1800, extracted from Thomas Milne's land use map by G.B.G. Bull.

Table 2. Acreage of and sales from gardens around London, 1798 (Middleton).

	Acreage	*Yield per acre*	*Total*
Neat Houses	200	£200	£40,000
Surrey side of the Thames	500	£150	£75,000
The out-skirts of London	1300	£100	£130,000
Wholly cultivated by the spade	*2000*	*£120 10s*	*£245,000*
Farming gardeners; their land partly cultivated by the spade, but mostly by the plough	8000	£50	£400,000

imported and varieties bred. Brompton Park was London's most famous early nursery and was located near a fashionable residential area. It was established in 1681 and within thirty years covered 50 acres. Many others were to follow its example of shadowing western residential development. Isleworth and Twickenham, as we have seen, were long famed for their tree nurseries.

By 1798, Middleton estimated that Middlesex alone had 1500 acres of nursery ground, yielding a gross income of £100,000 a year. He wrote of the sophisticated, international trade:

> At Chelsea, Brompton, Kensington, Hackney, Dalston, Bow, and Mile-end, much ground is occupied by nurserymen, who spare no expense in collecting the choicest sort, and greatest variety, of fruit-trees, and ornamental shrubs and flowers, from every quarter of the globe; and which they cultivate in a high degree of perfection....Many of them are annually exported to Ireland, Spain, Portugal, Italy, Russia, and, until lately, to France; but there are still greater quantities sold for use in this country.

Nurserymen were the aristocrats of the gardening community. If their businesses thrived, they were rich enough to become gentlemen. Their knowledge of gardening was such that they frequently wrote books on the subject and were the best botanists and plant breeders of their day. The nature of their trade meant they came into frequent contact with nobles and gentlemen who were interested in gardening and many had been head gardeners in private service earlier in their careers.[1]

1. The study of nursery gardening has for many years been dominated by the late John Harvey. His pioneering book, *Early Nurserymen*, published in 1974, is the best introduction to the subject, supplemented by his many other articles.

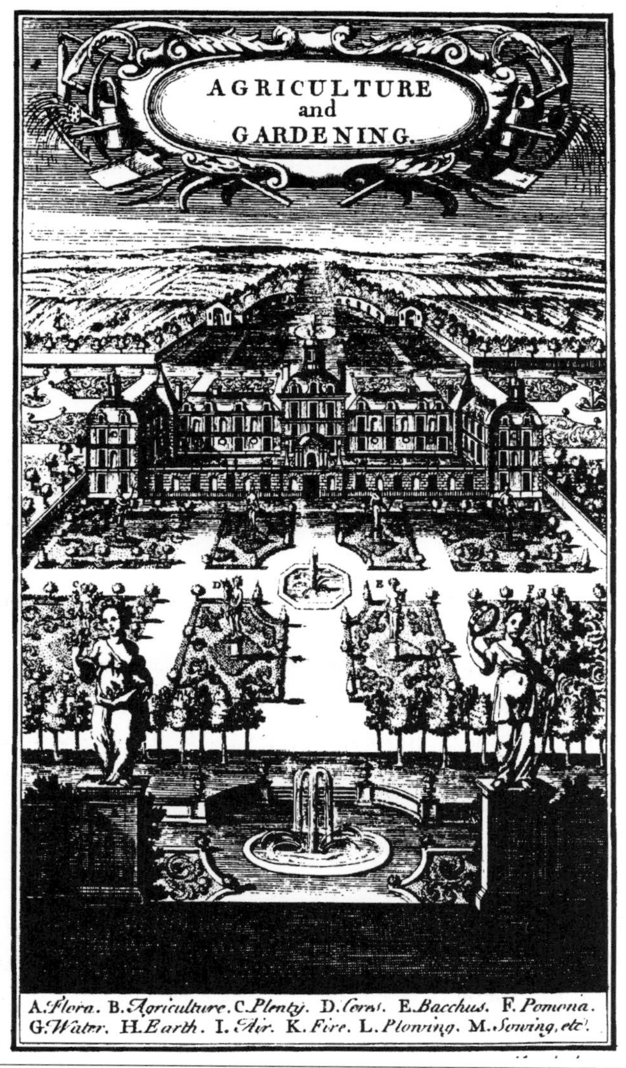

Figure 12. Agriculture and gardening symbolically linked in the engraved frontispiece of John Mortimer's *Whole Art of Husbandry* (1716). A gentleman's seat is set among gardens and fields, classical statutes representing agriculture and horticulture adorn the foreground and implements of husbandry and gardening surround the cartouche.

CHAPTER III

The spread of best practice

'The Gardeners about London have a dextrous and expeditious Way of doing with an Hough...'

London kitchen gardeners helped to spread horticultural and ultimately agricultural knowledge to gentlemen proprietors and, through them, to farmers. The Neat House gardeners had closer contact than many London colleagues with the gentry and interested amateurs because some of their gardens were open to inspection. Richard Bradley and Stephen Switzer went there when researching books on horticulture, and in 1798 John Middleton chose the Neat Houses as an example of the best practice in his survey of Middlesex (although they were outside the county, in Westminster). In garden literature of the later seventeenth and eighteenth centuries, London commercial kitchen gardeners as a whole were held up as paradigms for the gentry to emulate, and in books on agriculture one can also detect the influence of commercial gardening techniques.

Gardeners, however, produced and sold vegetables for many years before the gentry became interested in their skills. From the third quarter of the sixteenth century there are a number of recorded comments about the activities of commercial vegetable gardeners in England. Most are about gardeners in the London suburbs, although reference is also made to the refugees from the Low Countries in Norwich, Sandwich, Colchester and other towns in southern England before 1600. Until the 1660s, those who wrote on this subject praised the industry of the gardeners; were amazed at the output achieved from small acreages; saw gardening as a way of feeding the growing urban and rural poor and of avoiding famine in years of poor grain harvests; and recognized the high rents which landlords could ask for good garden land. None, however, recommended that commercial gardeners' techniques should be employed by the gentry in their kitchen gardens, although the expertise of some early nurserymen in raising fruit trees was signalled. This situation changed after the 1650s and for the next century and a half the skill of commercial producers was frequently extolled.[1]

The change is first evident in Robert Sharrock's book on the propagation of vegetables, published in 1660. He claimed he had written 'according to Observations made from Experience and Practice,' and recommends vegetable-growing techniques used by 'London Gardiners'. Stephen Blake, in a gardening book of 1664, also borrowed ideas from 'London Gardeners' and John Worlidge, in 1675, recorded vegetable growing practices employed by 'Gardiners near London'.[2]

For subsequent writers such as Bradley, Switzer and Philip Miller, such references were commonplace. Bradley mentioned individual gardeners by name. London practice continued to influence garden writers until the end of the eighteenth century but the initiative thereafter seems to shift towards innovations on the great country estates, either by proprietors or their gardeners. This is exemplified by J.C. Loudon's comprehensive *Encyclopaedia of Gardening* of 1824.

1. *Agricultural Change, Chapters from The Agrarian History of England and Wales 1500–1750*, ed. Joan Thirsk, 1990, vol. III, pp. 233–262; Hartlib, op. cit., pp. 8–9.
2. Robert Sharrock, *The History of the Propagation of Vegetables*, Oxford, 1660, pp. 15, 24; Stephen Blake, *Complete Gardeners Practice*, 1664, p. 130; John Worlidge, *Systema Agriculturae*, 1675, p. 33. From this period onwards, writers of gardening books who referred to 'London gardeners', 'kitchen gardeners', and simply 'the gardeners' in the context of a vegetable-growing technique employed by these groups which readers were urged to try, meant market gardeners. Usually this can be inferred from the context, and specific references in such terms to commercial production, as can be seen from some of the quotations below, provide confirmation.

The techniques which gardening writers thought worthy of translation from market gardens were many and various. Some can be seen as purely good ideas, 'gardening hints' such as those passed on by Philip Miller his *Gardeners Dictionary* or his *Kalendar*:

> There are many people, who are very fond of watering Cauliflower plants in summer; but the gardeners near London have almost wholly laid aside this practice, as finding a deal of trouble and charge to little purpose.[1]

> The Kitchen Gardeners (especially near London), have experienced, that by treating most of the esculent vegetables in a less tender manner, than was before practised, their crops succeed much better.

> [Welsh Onions] resist the severest frost...which is the occasion for their being much cultivated in gardens near London, some years since.[2]

The recommendations that stemmed from commercial practice reflected both the perceived strengths of the growers and also what an author believed would interest his readership. Occasional mention was made of ways of protecting against wind and cold, such as the use of reed hedges or ridging; there were some hints on pest control; advice was given on particular crops in favour with the rich: asparagus, melons, and kidney beans; and significant space was devoted to methods of producing vegetables out of season using hotbeds and glass.

Philip Miller prefaced full directions on raising early peas with the comment: 'It is common practise with the gardeners near London, to raise Peas upon hot-beds, to have them very early in the spring.' John Abercrombie announced that: 'Many of the kitchen-gardeners about London begin to make asparagus hot-beds about the middle or latter end of September, or early October, in order to have

1. Philip Miller, *Gardeners Dictionary*, 1763.
2. Philip Miller, *The Gardeners Kalendar*, 1765, Preface, pp. 144, 304.

asparagus fit to gather by Lord Mayor's Day, which mostly happens the second week in November.' Methods of raising of other vegetables on hotbeds were described. In 1714, John Laurence rather haughtily dismissed much kitchen garden practice as too well known to bother with, but he did concede the importance of hot-beds and the pre-eminence of London gardeners in the matter, 'Only there is one thing relating to the Management of Hot-Beds, whereon Mellons, Cucumbers, etc. are wont to be raised, which it may not be amiss here to take notice of; because, tho' it hath been practised with Success by some of the Gardiners near London; yet other Persons curious in that Matter, not being apprized of it, may think themselves obliged for the Relation.'[1]

Readers of gardening books were also advised to copy London gardeners in rotating crops, manuring heavily, digging, hoeing and generally stirring the ground frequently. The comments of Leonard Meager in the 1670s make it clear that London gardeners were considered exceptional in these matters, the intensity of their cultivations exempting them from the usual problems of monoculture.

> Another thing I would have you take notice of, and that is, that you do not sow one sort of crop too often upon one and the same piece of ground, but sow it with changeable Crops, especially Parsnips and Carrots, the which being sown too often without change, will be apt to canker, rot, or be very apt to be worm-eaten, although the ground be maintained very rich. I do not speak this of the great Garden-grounds in or near London, where their grounds are in a manner made new and fresh once in two or three years, by dung and soil, and good trenching; so that their ground is as it were new and fresh for one and the same kind of Crops every year.[2]

1. Worlidge, *Systema Agriculturae*, 1675, p. 33; Miller, *Gardeners Dictionary*; John Laurence, *A New System of Agriculture*, 1726, p. 373; William Ellis, *Agriculture Improv'd*, 1746, vol. II, p. 64, Thomas Mawe & John Abercrombie, *Every Man His Own Gardener*, 1813, pp. 373, 587; John Laurence, *The Gentleman's Recreation*, 1714, p. 74.
2. Leonard Meager, *English Gardener*, 1670, p. 165.

Not only were there no fallows in London gardens, but there was multiple cropping, either sowing mixed seed at the same time on a bed or growing different crops side by side in rows, with each crop harvested separately. Robert Sharrock found the first method in London gardens during the 1660s, ' Many times they soe divers seeds in a Bed together, as Radishes and carrots, that by such time as the Carrots come up, the Radishes may be gone...London Gardiners sow Radish, Lettice, Parsley, Carrots, on the same bed, gathering each in their seasons,' Mixed beds were still being recommended as a London gardeners' practice in the middle of the next century: 'The red Beet is frequently sown with Carrots, Parsneps, or Onions, by the kitchen gardeners near London, who draw up their Carrots or Onions when they are young, whereby the Beets will have room to grow,...indeed, the gardeners near London mix Spinage with their Radish seed, and so have a double crop; which is an advantage where ground is dear.'[1]

A sophisticated variant of this technique was to plant or sow in rows set widely apart with a different crop between them, or to have even three or more crops alternating in rows. Hence Philip Miller's advice on cabbages: 'Common white, red, flat, and long-sided Cabbages are transplanted (which in the kitchen gardens near London, is commonly between Cauliflowers, Artichokes, &c at about two feet and a half distance in the rows).' His suggestion for cucumbers was, 'if you intend to make ridges for Cucumbers between the rows of Cauliflower plants (as is generally practised by the gardeners near London), you must then make your rows eight feet asunder.'

Sir John Hill neatly displays the links between commercial and private gardens: 'In Places where Cabbages are raised for sale, Ground may be spared by planting them between the Rows of other Crops, as Artichokes, and the like; this may be learn'd in every Gardeners' Ground about London: or where there is want of Room, a Practice of that kind must be admitted elsewhere.'[2]

1. Sharrock, op. cit., 1660, pp. 14–15; Miller, *Gardeners Dictionary*, 1763.
2. Miller, ibid.; Sir John Hill, *Eden*, 1757, p. 286.

London gardeners thinned their rows of young vegetables, especially those, such as root crops, which needed room to swell. In 1664, Stephen Blake recommended following London methods when growing carrots, 'That you let them not grow too thick...the best way to prevent this, is to hoe them as our London Gardeners do, so that each Carrot stand ten inches one from another, or thereabout.' John Laurence found the technique of hoeing turnips particularly interesting in the 1720s: 'They must be carefully houghed and sized at two different times, which the Gardeners about London have a dextrous and expeditious Way of doing with an Hough about six inches wide.'[1]

Where a book of instruction in kitchen gardening was written by an author with commercial experience, it is no surprise that he drew on his own practices for inspiration. But other writers were landowners or members of the gentry and they too used London as their models. It was generally believed that city gardeners were better at their craft than both gentlemen and their gardening employees. They were so much better that there was a tendency to consider any technique used in London kitchen gardens one that private gardeners should follow. Occasional mention of this error was made by Philip Miller but the most forceful rejection of slavish adherence to London practice was voiced by Sir John Hill. He was particularly concerned that gentlemen who were not constrained by space and high rents would needlessly follow the London gardeners in multiple cropping. In a book published in 1757, he repeatedly advises against this:

> One common Error we are here to caution the Gardener to avoid, which is the sowing the Ground with another Crop whereon these cabbages are to be planted.
>
> This is a common Practice about London, where Land is very dear, and the People who raise these Things must make

1. Blake, op. cit., p. 130; John Laurence, *A New System of Agriculture*, 1726, p. 367.

> the most of every Inch of it: and from their Practice it is transcribed into the Books upon this Subject.
>
> The usual Crop is Spinach, and they sow it before they plant the Cabbages; but in this Case both are the worse for it, and there is no Reason why the Country Gentleman, who has Ground enough should impair two Crops, by dividing the Nourishment, which is only sufficient for bringing one of them to Perfection.[1]

Confirmation of this outside interest in London commercial gardening is shown in the letters written by the landowner John Cockburn from the London suburb of Tottenham during the 1730s to his gardener, Charles, at home in Ormiston, near Edinburgh. He gave advice on vegetable production in his own garden and commercial production on Charles' father's land. He lectured on the need to educate the taste of Edinburgh consumers, 'and if you can once get into the custom of some who have it [i.e. taste], will put others upon enquiring where they had good things'. He suggested sending 'a dish of young pease or Beans to any of your Customers when only old are to be had,' in the hope they would tell their friends about it. Charles was advised to have crops out of season: 'in winter what is proper then as in Summer or when everybody has not an over Stock of the Same things'. His employer reminded him that 'Cuff', a Tottenham market gardener under whom Charles had served, 'took care to have something for the market every day the year round.'[2]

Cockburn questioned a local kitchen gardener who came to his door selling vegetables in June 1735, reporting to Charles his prices, the rents and tithes the man paid, and his costs of labour and dung, remarking that if London gardeners could make profits with such

1. Malcolm Thick, '"Superior Vegetables", greens and roots in London, 1660–1750', in *Food, Culture and History: Papers of the London Food Seminar*, vol. I, ed. G. & V. Mars, 1993; Sir John Hill, *Eden*, 1757, p. 108.
2. *Letters of John Cockburn of Ormistoun to his Gardener, 1727–1744*, ed. James Colville, Scottish History Society, XLV, 1904, Edinburgh, pp. 17–19, 25, 28.

high overheads, there was scope to increase production around Edinburgh. Charles was also given 'tips' gleaned from London gardeners: how to transport produce in baskets; trenching and ridging against winter cold; how to soften water to be used on cauliflowers. In an echo of arrangements seen at the Neat House gardens, Cockburn advised him to open 'a good public house' next to his father's market garden, with good liquor and where the gentry could be served such delicacies as a dish of soft fruit in season.[1]

If commercial gardeners influenced gentlemen when they considered their own vegetable gardens, did gardening, in turn, impinge on their attitudes towards farming? Joan Thirsk has discussed the interaction of gardening with farming from the beginning of the seventeenth century until 1750, showing a movement both of garden crops and techniques from garden to field. She has also put these trends in the context of an agrarian economy searching for new crops and methods of production in the face of depressed markets for grain. I will, to some extent, cover this ground again, but wish to examine further the possible influence of kitchen gardening practices on gentlemen interested in agricultural improvement, specifically by looking for links in contemporary literature on agricultural improvement.[2]

The connection between gardening and farm husbandry is sometimes implicit in discussions of garden crops and techniques to be tried in the fields. Jethro Tull rarely mentions gardening in *Horse-hoing Husbandry* although, essentially, he is proposing mechanized gardening techniques. Some may have agreed with Richard Weston in 1773 that, 'It may seem odd to some, that I should propose the entering so much into the plan of kitchen-garden crops, since these are in the general opinion rather beneath the notice of a gentleman'. Weston was, however, firmly convinced of the utility of talking to gardeners and, 'consulted and tried the methods practised by the

1. Ibid., pp. 26, 28–29, 50–1.
2. Thirsk, *Agricultural Change*, *ut sup.*, pp. 260–2, 311–317.

most skilful nurserymen, and London kitchen-gardeners,' concluding, 'if a gentleman hoped to reap any benefit from his labour, it must be by uniting the garden-culture with farming, and changing the common modes of cultivation.' Edward Lisle talked to gardeners when compiling his *Observations in Husbandry* (1757), and the numerous books by Richard Bradley earlier in the same century demonstrate the keen interest some gentlemen took in London commercial gardening and its possible application to farming.[1]

Some writers made clear the general links between gardening and agriculture: the sixteenth-century German writer Conrad Heresbach, as translated by Barnaby Googe, followed a chapter on arable production with one on gardening and orchards, 'because of the alliance betwixt Hearbes, Trees, and Corne, and because their husbandry is almost one'. In 1718, Stephen Switzer wrote of 'Agriculture (with which Gard'ning is inextricably wove)'.[2]

The possible benefits of digging instead of ploughing were debated over many years. In 1600, Sir Hugh Platt was convinced of the advantages of digging: 'When and how to digge, weed, or trench your grounds with the spade, is a matter so trivial and well knowne already to everie countrie Coridin [i.e. rustic],...onely the depth of digging, and true laying of the ground, seemes to be materiall in this our new kind of husbandry.'[3]

Samuel Hartlib and his circle continued the debate and the consensus in the mid-seventeenth century was that digging was a technique well worth investigating. By 1675, John Worlidge concluded the plough still had the advantage over the spade, on technical as well as cost grounds, but John Laurence gave the spade a more sympathetic response in 1726:

1. Richard Weston, *Tracts on Practical Agriculture and gardening*, 1773, p. xvii; Edward Lisle, *Observations in husbandry*, 1757, vol. II, pp. 271–284.
2. Gervaise Markham, *Art of Husbandry*, 1631, p. 87; Stephen Switzer, *Ichnographia Rustica*, 1718, p. vi.
3. Sir Hugh Platt, *The New and Admirable Arte of setting Corne*, 1600.

> Some are of Opinion, that Digging of Land is preferable to the Plough: for by that Means it is all deep enough, and you may have a great deal of fresh Mould, and can without Clods have it as fine and smooth as you will, neither is the Expense of it so great as some Persons may think, it being done about London (as I am told) for Two pence a Rod, or four Nobles the Acre; but I do not suppose it to be new Ground.
>
> If it be this; what is the extraordinary Charge of Digging, etc. may well-nigh be saved in the Seed; and it may be managed so as better to be preserved from Birds or Vermin, and then you have the Goodness of the Soil into the Bargain. I have seen a great deal of Digging for Beans and Roots in Common Fields, on both Sides of Ebbisham in Surrey, and other Places in Berkshire; but yet I find the Gardeners near London love the Plough, and when their Land is well ploughed, they employ Men to hoe little cross Furrows to lay Beans, or trill Pease in, and so with the Hoe again cover them. And why may not it be thus with Wheat?
>
> The Hoe at all times will clear the Weeds and supply the Grain with good Earth. If Land be shallow, I presume it may be worth while to lay one Part of the Land upon the other, and there set; which, in small Quantities, is easily tried.

Within a few years Jethro Tull produced his book on horse-hoeing, explaining that he had found hand cultivation on his farm impossible because of truculent labourers. Digging dropped out of the agricultural debate but discussion of other garden-based innovations such as seed-drilling, row cultivation, and hoeing, continued.[1]

Sir Hugh Platt opened the discussion of setting grain—dropping individual seeds into holes at set distances—in a pamphlet of 1600. He linked the first trials with gardening, imagining how the idea first occurred:

> Heere I maie rather probablie coniecture then certainly determine howe this new conceit in setting of graine began.

1. John Laurence, op. cit., pp. 90–91; Hartlib, op. cit., pp. 6–8; Worlidge, *Systema Agriculturae*, 1675, pp. 34–5, Jethro Tull, *The Horse-hoing Husbandry, or an Essay on the Principles of Tillage and Vegetation*, 1733.

> Happily some silly wench having a fewe cornes of wheate, mixed with some other seed, and being carelesse of the worke shee had in hand, might nowe and then in steed of a Raddish or carret seede, let fall a wheate corne into the ground, which after braunching it selfe into many eares, and yeelding so great encrease, gave iust occasion of some farther triall. Peradvanture the great and rich fertilitie that doth usually happen in the setting of beanes and pease, might stirre up some practising wit or other to make the like experience in wheate and barley.[1]

He, and Edward Maxey in 1601, advocated using a hand-held setting board, keeping to rows controlled by a garden line. Gabriel Plattes in 1639 also advanced the cause of setting but, by 1655, although setting of beans and peas was thought a good idea, setting grain was not:

> 1. Because to set Corn is an infinite trouble and charge; and if it be not very exactly done, which children neither can nor will do, and these must be the chief setters; will be very prejudiciuous.
> 2. If worms, frost, ill weather, or fowls, destroy any part of your seed, which they will do; your crop is much impaired.
> 3. The ground cannot be so well weeded, and the mould raised about the roots by the how. Which 3 inconveniences are remedied by the other way.
>
> Further, I dare affirm, that after the ground is digged or ploughed and harrowed: even it's better to howe Wheat in, then to sowe it after the common Way; because the weeds may be easily destroyed by running the howe through it in the Spring, and the mould raised about the roots of the Corn, as the Gardiners do with Pease, it would save much Corn in dear years, and for other Reasons before mentioned. Yea, it is not more chargeable; for a Gardiner will howe in an Acre for 5s and after in the Spring for less money run it over with a howe, and cut up all the weeds, and raise the mould: which charges are not great, and you shall save above a bushel of seed, which in dear years is more worth then all your charges.

1. Platt, op. cit.

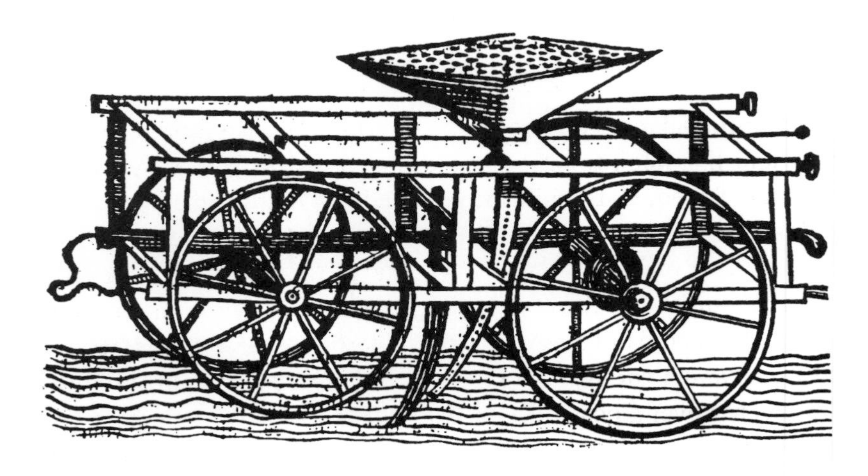

Figure 13. John Worlidge's seed drill, 1675; an early attempt at mechanized row cultivation in fields.

It is significant that the alternative to setting, hoeing-in, was a commercial gardening practice which the author, a correspondent of Samuel Hartlib, knew well.[1]

Hartlib's informant also knew that London husbandmen who grew vegetables were using another method of sowing in rows, the drill plough, for he continued that anyone not wishing to hoe-in seeds, 'let him use a Dril-Plough, with one horse, which is commonly known at Fulham, and about London'. Joan Thirsk has discovered that many gentlemen were experimenting with drill ploughs in the second half of the seventeenth century, and they were in use in Surrey near London. John Worlidge published the specifications of a seed-drill he had invented in 1675 (figure 13), to be followed by the famous experiments of Jethro Tull in the 1730s, another drill made by William Ellis in the 1740s and by many more attempts to

1. Gabriel Plattes, *Discovery of Infinite Treasure*, 1639, pp. 49–58; Hartlib, op. cit., pp. 6–8; Platt, op. cit.; E. Maxey, *A new instruction of plowing and setting of corne*, 1601.

produce workable machines thereafter. The basic methods of kitchen gardeners, sowing by drilling and cleaning crops by hoeing between the rows, were at the root of these experiments.[1]

Economic adversity after the Restoration led farmers to try new crops and livestock to supplement their incomes. Writers, such as Adolphus Speed, Gabriel Plattes and Samuel Hartlib, recommended gardening and, more particularly, production of garden crops on farms. Whilst the more fanciful recommendations, such as Speed's 500 acre estate with 20 acres of liquorice, 20 acres of pumpkins and cabbages, 20 acres of French beans, 5 acres of clove gilley flowers and roses, and a variety of other garden produce grown on a similar scale, were never a possibility, Joan Thirsk has shown that there was significant production of such 'alternative' crops. In areas of mixed farming around London, and in favourable areas some distance from the capital, such as Sandwich in Kent or Sandy in Bedfordshire, husbandmen incorporated vegetable production into their farms and alternated corn and other conventional crops with vegetables produced with the plough or spade, becoming farmer-gardeners.[2]

The main roots grown commercially by gardeners—turnips, carrots, and parsnips—were also seen, by farmers, as potential field crops for animal fodder. Only the turnip was to be of lasting importance. The timescale of the spread of turnips as a fodder crop has been a matter for debate but, by the early eighteenth century, agricultural writers recognized it as a field, rather than garden crop. 'The Turnep has been formerly thought to be a Root only fit for the Garden and Kitchen Use; but the industrious Farmer, finds it now to be one of his chief Treasures,' wrote John Laurence in 1726. William Ellis, in 1747, claimed that, 'this root is become very much in use in divers countries in this kingdom, and become the food of beast and sheep etc.'[2]

1. Hartlib, op. cit., pp. 6–8; G.R. Fussell, *The Farmer's Tools*, 1981, pp. 92–115.
2. Thirsk, op. cit., pp. 263–319; A. Speed, *Adam out of Eden*, 1659, pp. 142.
3. John Laurence, op. cit., p. 109; William Ellis, *The Farmer's Instructor*, 1747, p. 63.

Tull, and his successors, recommended drilling and hoeing turnips and it is clear that the farmers who adopted them as a field crop in the eighteenth century looked to gardeners for guidance in their cultivation. This was clearly described by Philip Miller in 1747:

> The general method now practised in England, for cultivating this plant in the fields, is the same as is practised by the farming gardeners, who supply the London markets with these roots, and is the same as before directed. But it is only within the compass of a few years, that the country-people have been acquainted with the method of hoeing them; so that the farmers formerly employed gardeners, who had been bred up in the kitchen-gardens, to perform this work. The usual price given per acre, for twice hoeing and leaving the crop clean, and the plants set out properly, was seven shillings; at which price the gardeners could get so much per week, as to make it worth their while to leave their habitations, and practise this in different counties, during the season for this work, which always happens, after the greatest hurry of business in the kitchen-gardens is over; so that they usually formed themselves in small gangs of six or seven persons, and set out on their different routs, each gang fixing at a distance from the rest, and undertaking the work of as many farmers in the neighbourhood, as they could manage in the season; but as this work is now performed by many country labourers, that practice is lost to the kitchen-gardeners, the labourers doing it much cheaper.[1]

Contemporaries and historians have recognized the role of the gentry in promoting innovations in agriculture and horticulture both by writing about new techniques and crops and by doing practical experiments. John Worlidge, in 1675, acknowledged that vegetable growing was more risky than more conventional arable production and husbandmen might be discouraged by a bad harvest: 'But we hope better of the Ingenious, that they will set to their helping hand to promote this useful and necessary Art, and thereby become a

48. Miller, *Gardeners Dictionary*, 1763.

provoking President to their ignorant Neighbours'. A century later Richard Weston was convinced, 'that husbandry will never arrive at half the perfection that it is capable of, till the garden-culture is more imitated in the field'. William Marshall in 1799 was slightly more cautious but he had tried gardening techniques on his own farm: 'It would not, perhaps, in the present state of things, be eligible to reduce each field to a garden; but I would wish to see an arable field and a kitchen garden bear some resemblance.' Although the gentry were interested in moving the garden into the field, most farmers were not convinced until well into the nineteenth century of the merits of the changes they advocated because hard economics did not justify their adoption.[1]

The writers who thought gardening had lessons for agriculture were, from the outset, intrigued by the farmer-gardeners near London. Farm-gardening declined after 1750 when grain prices picked up, but in many parts of Surrey and Middlesex such farms continued to supply London into the Victorian era. In Chelsea and Fulham, farmer-gardeners occupied much of the land from the early seventeenth to the middle of the nineteenth centuries. These 'hybrid husbandmen' were seen as living demonstrations of the way in which agriculture could learn from gardening. They dispensed with fallows; added dung and other fertilizers in large quantities to their soil; mixed digging and ploughing, stirring the land at greater depth than most farmers; sowed seed in rows, often by mechanical means; kept down weeds by hoeing and hoed to thin crops where applicable; and grew roots and other garden crops in fields. The agricultural journalist John Houghton wrote: 'If I were worthy to advise, I would have the country farmers send their sons they design to breed in their own way, to live a year or two with the husbandmen about London, that are partly gardiners and partly ploughmen, or at least to take

1. Worlidge, op. cit., pp. 132–3; Richard Weston, op. cit., pp. x–xi; William Marshall, *Minutes etc. on Agriculture in the Southern Counties*, 1799, vol. I, pp. 173–4.

some servants that have been bred to.' His advice may have fallen on many deaf ears at the time but in due course most farmers adopted techniques used by the farmer-gardeners of the London area.[1]

Joan Thirsk, surveying the period 1640 to 1750, and looking forward to the nineteenth century, summed up the interaction thus:

> For many decades—almost a century, in fact—the success of horticulture proffered lessons to agriculture, and they did not pass unheeded. Many writers commented on the high productivity of land that was dug with the spade, and the fears of some gentry that soils would be spoiled by so much digging were generally dissipated. Then in the early eighteenth century, men who positively hankered for the more meticulous cultivation of arable land by methods akin to those of the gardeners added a further consideration to their reflections, namely the rising cost of labour. In this conjuncture of circumstances there developed a more positive interest in using wheeled implements in the fields for drilling and hoeing. In the more encouraging circumstances after 1750 this was to usher in a revolution in techniques of cultivation.[2]

1. Thomas Faulkner, *History and Topography of Chelsea & Environs*, 1810, p. 14; John Houghton, *A collection for Improvement of Husbandry and Trade*, 1727, vol. III, pp. 233–5.
2. Thirsk, op. cit., p. 311.

II

THE NEAT HOUSE GARDENS

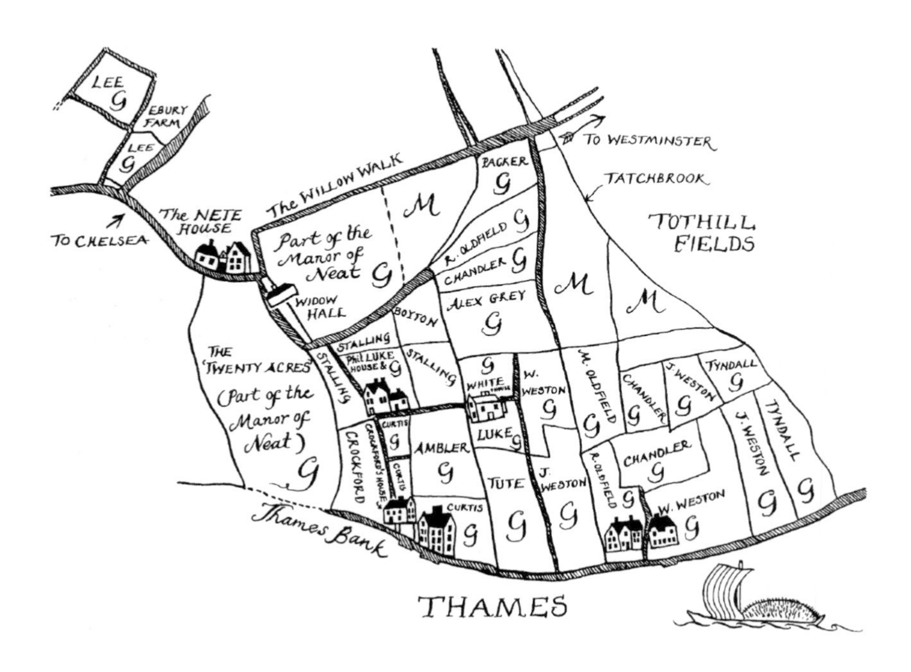

Figure 14. The southern half of the estate of Mary Davies, acquired by marriage by the Grosvenor family in 1677, from a survey of 1675. 'G' denotes market garden; 'M' indicates meadow.

CHAPTER IV
Origins

'All that site, soil, circuit, and precinct of the manor of Nete...'

The Neat House gardens were just a small part of the market gardening which surrounded London. They existed for over two hundred years, from about 1600 until the second quarter of the nineteenth century. What follows is a history of the gardens, primarily social and economic, but written for anyone with a general interest in the history of gardening. Most of the material on which this study is based dates from the late seventeenth and early eighteenth centuries and this period, in consequence, receives most attention.[1]

The area once occupied by the gardens is today an inner suburb of London, a part of Westminster known as Pimlico. Their southern boundary in about 1700 was the Thames, curving in a long bend from the present railway bridge which takes the lines into Victoria, eastwards almost to today's Vauxhall Bridge. The eastern boundary is now traced inland along Tachbrook Street (built over the course of the brook) as far north as Warwick Way (anciently the Willow Walk from Westminster to the Neat House). Warwick Way marks the northern boundary and we follow this road westwards to Ebury Bridge then turn south by the railway lines along the western edge back to the Thames.

1. The main sources of information used are the archives of the Grosvenor Estate, mostly deposited at the City of Westminster Archives Collection; other archives in this library relating to Westminster; wills and inventories at the Greater London Record Office; secondary and other sources cited in footnotes.

The gardens were mostly in the southern part of the manor of Ebury, with some in the small sub-manor of the Neat, itself an enclave encircled by the larger jurisdiction (see figure 14). Both were in the south-west corner of the broad parish of St Martin-in-the-Fields. When that was divided in the eighteenth century, they fell into the new parish of St George, Hanover Square. Once on the edge of the city, St Martin and the new parishes created from it were engulfed by the tide of London, the Neat Houses holding out long after neighbouring areas had disappeared under buildings and roads.

For a long time, this corner of the parish was isolated and unimportant; the only major road, between Westminster and Chelsea, ran just north of the Neat Houses. Its relative peacefulness was recognized by the abbots of Westminster, who owned the whole area until the 1530s. They occupied the moated house of the sub-manor of the Neat (the 'Neat House') as a rural retreat.

This was about a mile from Westminster Abbey, some 450 yards away from the Thames bank, at the end of the Willow Walk. For centuries the abbots enjoyed gardens within the two acres enclosed by the moat.[1] An historian who found many references to these in the Abbey archives described the Neat House thus: 'It had five gardens, for which the abbot's gardener bought lettuce, savory, borage, chervil and violet seeds in 1327. Besides a herb garden it had turfed alleyways and walks, and Westminster Abbey accounts are full of references to the Abbot's stays at 'La Neyte' and of the King's visits there. The King's gardener from 'le paleys' [the palace] went there in 1345 to cut some small willow twigs to tie up the palace vines, from what was obviously a willow plantation, part of a nursery and vegetable patch containing willows, flax, hemp and peas.' The 1327 purchase also included onion, leek, parsley, beet, orach, spinach, cresses, worts, hyssop, fennel, coriander, *langedbefe*, and hemp. In

1. Charles T. Gatty, *Mary Davies and the Manor of Ebury*, 1921, vol. II, plate 31; W. Loftie Rutton, 'The Manor of Eia, or Eye next Westminster', *Archaeologia*, vol. LXII, 1910, p. 38.

1275–6, the garden was growing skirrets. The fourteenth-century Neat House gardens must have provided a range of produce for the household. The gardens were still there when the manor was ceded to the Crown in July 1536: the deed of surrender included 'All that site, soil, circuit, and precinct of the Manor of Nete within the compass of the moat, with all the houses, buildings, yards, gardens, orchards, fishings, and other commodities in and about the same site.'[1]

Could these moated gardens be the origin of market gardening at the Neat Houses? Medieval monasteries, and other large households, sold surplus garden produce in local markets. Some monasteries, including Westminster Abbey, leased all or part of their gardens to commercial gardeners in return for rent in money or kind. If this happened at the Neat House it would ensure that the garden continued as a commercial venture during the upheaval of surrender to the Crown. It appears, however, that the gardens within the moat were not sublet. A royal survey in 1536, prior to the surrender, found that the Abbot derived no rent from the Neat House.

The Crown for years allowed various people to use the Neat House manor house as a 'grace and favour' dwelling before it was eventually granted to a courtier. He sublet it, and it was described as a farm housing one man in 1549; hardly a description of a market garden. Gardens do not take kindly to neglect and it is doubtful if they would have been viable in the 1540s unless they had been constantly tended since 1537. In any case, market gardens much closer to London and Westminster supplied vegetables in the mid-sixteenth century, making any at the relatively isolated Neat House uneconomic. We must conclude that the Neat House—which eventually became a ruin, and may, in the seventeenth century, have been a roadside pleasure garden and tavern—gave nothing but its name to

1. Loftie Rutton, op. cit., pp. 38, 56; Theresa McLean, *Medieval English Gardens*, 1981, p. 108; John Harvey, *Mediaeval Gardens*, 1981, p. 154; Chester City Records Office, Grosvenor Records, Box 77/2.

the gardens. The origin of commercial gardening is more likely found in the upheavals in the parish caused both by the dissolution of the Abbey and general economic change.[1]

The Crown used the former monastic lands in Westminster to pay courtiers and Crown servants, splitting the estate into sections which, in their turn, were often sublet or resold. The manor of Ebury was granted to John and Isabell Wevant for 41 years from Michaelmas 1543. In 1567 the reversion of this lease (from Michaelmas 1584) was granted to William Gibbes for 31 years and he sold it within a week. In 1585 a reversion was granted to another courtier, Sir Thomas Knevett, to run from 1615 for 60 years. In 1591 this same reversion was granted to Sir Humphry Lynde and Edmond Doubleday. Lynde's son and Doubleday formally divided the manor between them in 1614. Lynde junior's moiety had passed by sale through three other hands by 1620. In 1623 the freehold of the manor, subject to the leases, was sold by the Crown to Sir Lionel Cranfield who had already, in 1620, acquired Lynde's leasehold share. Finally, the prosperous London scrivener Hugh Audley purchased the freehold from Cranfield in 1626, subject only to the lease granted originally to Doubleday on half the estate. This lease expired in 1675, leaving the young heiress Mary Davies, who inherited Audley's estate, in control of the whole manor. On her marriage in 1677 to Sir Thomas Grosvenor, the land was acquired by the family which still holds it today. Below the level of these freeholders and head-lessees, further subletting occurred before tenure reached the farming and gardening tenants.[2]

Such fragmentation turned land into a commodity, to be bought and sold for profit or investment, with rents set as high as the market

1. City of Westminster Archives Collection (henceforth C of WAC), 1040/10/Box 32/1; Gatty, op. cit., vol. I, p. 38 ; John McMaster, *A short history of the Royal parish of St Martin-in-the-Field*s, 1916, p. 256; Gervaise Rosser, *Medieval Westminster, 200–1540*, 1989, pp. 134–7; John Harvey, *Early Nurserymen*, pp. 39–41; John Stow, *A Survey of London*, ed. Morley, p. 151.
2. Gatty, op. cit., vol. I, pp. 42–5; McMaster, op. cit., p. 258.

would bear. In 1549 a petition to the Crown claimed that rents in the area had greatly increased and that Sir Anthony Browne let farmland in the manor of Neat 'to them that will geve most mony ffor ev'ry Acre'. The petitioners' main grievance was the extinction of Lammas grazing rights following enclosure of whole farms by the Crown's subtenants and the piecemeal enclosing of meadow and arable land. Associated with the new closes was a switch from arable to pasture 'and anchyant men saye that ther have byn vi plowes more than now ys to the Decay of husbandrye.' Riots and a petition in 1592 show the inhabitants of St Martin-in-the-Fields still fighting to regain their Lammas grazing rights in the face of more and more enclosure by the unrestrained subtenants.[1]

A complaint in 1549 that thieves and harlots were more frequent in the parish highlights the physical encroachment by the poorer outer suburbs as well as by fashionable overspill from St James's and Westminster. Enclosure, mainly for pasture, can be viewed as economic encroachment. Enterprising husbandmen, free from the customary constraints of common Lammas grazing, exploited the growing market for food in London and Westminster whilst their immediate landlords reaped the benefits of increased rents. Those who relied on customary grazing rights lost out.

One of the new ways of making money from this same demand was market gardening. Like cattle grazing, it was well suited to the recently enclosed fields. The origins of the Neat House gardens lie in this conjunction of demand and availability of suitable land.[2] The earliest identifiable market gardens at the Neat Houses date from the second decade of the seventeenth century, when commercial kitchen gardening expanded rapidly around London and Westminster, particularly on the western edge.

Anti-enclosure rioting in the parish in 1592 gave rise to petitions and court appearances, but none of the complainants mentioned

1. Gatty, op. cit., vol. I, pp. 64–7; McMaster, ibid.; C of WAC, F2001, fol. 2–3.
2. C of WAC, F2001, fol. 2–3.

enclosure for gardening. As such a novelty might well have merited notice, we may assume no gardening then existed. Twenty-one years later, in 1614, a map (now in bad condition) was made to accompany the division of the lease of the manor of Ebury between Doubleday and Lynde. It contains the first reference to market gardens in its survey of the lands attached to the sub-manor of the Neat. The field called 'Twenty Acres' (actually containing 17 acres and 30 perches) is described as 'The Gardens'. The use of the plural implies more than one holding. No other land was described as garden.[1]

A memorandum of 1622 relating to the southern half of the manor of Ebury lists two parcels of garden ground, 'One other peece by the Thames in garden' of nine acres, and 'One peece of Garden ground in the occupacion of ffrost at the rent of £24', apparently of eight acres. Frost's land is further identified in a map produced sometime between 1663 and 1670 to clarify ownership of the manor. It is the 'Close between Lane Meadow and the Thames' which abutted the 'Twenty Acres' to the east and quotes an earlier survey giving the area as 13 acres at a rent of £24. The nine-acre garden was a Thames-side field described in 1614 as 'One Meadow ioyning East'. Neither the surveys of 1622 or 1663–70 covered the 'Twenty Acres', but we may assume that gardening continued there.[2]

When Edmund Doubleday's lease of Ebury Manor expired in 1675, leaving the heir of Hugh Audley's estate in control of the whole of it, market gardening had been quietly expanding for more than fifty years. So far, the estate had had no dealings with the ultimate occupiers of the land and had only a sketchy idea of its use. When the lease fell in, a detailed survey was executed (figure 14). It found the majority of the occupiers were market gardeners, occupying ground totalling over 90 acres. The two large fields in the sub-manor of Neat, Twenty Acres and Cawsay Hall (south of Willow

1. Gatty, op. cit., vol. I, pp. 64–7, op. cit., vol. II, plate 31; McMaster, op. cit., pp. 257–9; C of WAC, 1049/12/115.
2. C of WAC, 1049/1/28; 1049/12/115; 'A Plan of Ebury Manor', London Topographical Society, 1915.

Walk), were not included in this survey, but were also occupied by gardeners. Thus the acreage certainly or probably under gardens rose from up to 17 in 1614, 39 in 1622, to at least 107 in 1675.[1]

Intensive vegetable gardening was a radical departure from traditional husbandry and we must consider who were the innovators. The survey in 1675, and subsequent leases and rentals provide the names and locations of most of the gardeners at the Neat Houses. Elsewhere, gardening, and related intensive agriculture, were introduced in differing ways by people of widely differing origins and rank. I conclude from an examination of surnames entered in the parish registers between 1550 and the 1630s, a selection of rating assessments from 1599 to 1664, and the early records of the Gardeners' Company that gardening was started at the Neat Houses in the 1600s largely, if not entirely, by incomers of English origin.[2]

Jasper Stalling and his son Lott are the only early gardeners at the Neat Houses whose names imply origins in the Low Countries or northern France. Jasper Stalling first appears as a ratepayer of St James's parish in 1617, the year of his admission to the Gardeners' Company. Jasper may well have been the son or a close relative of

1. Gatty, op. cit., vol. II, p. 190; C of WAC, 1049/1/3; 'A Plan of Ebury Manor', London Topographical Society, 1915 , map and introduction p. 2. This map, originally produced sometime between 1663 and 1670 for Audley's legatee, Alexander Davies, contains much detail on the tenurial history of the freehold part of the estate, but has far less information about the half still occupied by the lessees; C of WAC, 1049/12/114.

2. *A Register of Baptisms, Marriages and Burials in the parish of St Martin-in-the-Fields, 1550–1619, and 1619–1636*, Harleian Society, vol. 10, 1899; vol. 46, 1936; C of WAC, F3354; F4538; F325; F328; F335; F345–6; F348–9; F352; F391; Guildhall Library, MS21.128; MS3389/1a; MS3389/2; J.V. Kitto, *The Royal parish of St Martin-in-the-Fields, The Accounts of the Churchwardens, 1525–1603,* 1901, name index.

Comparing the 18 gardeners' surnames in the 1675 Grosvenor Estate rental with the names in St Martin-in-the-Fields parish registers between 1550 and 1636, we find 8 of the 1675 surnames between 1550 and 1619, and 11 in the period 1619 to 1636. Comparing surnames in the same rental with a selection of rating assessments for St Martin-in-the-Fields we discover 6 coincidences between 1599 and 1619, 10 between 1620 and 1626, and 14 in 1663 and 1664.

the immigrant William Stallenge, who was 'Comptroller of the custome house' to James I. William obtained a patent to import mulberry seed and, in 1609, together with a Frenchman called Verton, organized King James's scheme, based on ideas tried by Henri IV of France, to plant mulberries in most counties of England to encourage silkworm breeding and the silk industry. William Stallenge was a leading breeder of silkworms and had charge of the royal garden created on '4 acres of land taken for His Majesty's use, near to his Palace of Westminster, for the mulberry trees'. This mulberry garden was only a little north of the Neat Houses. Evidence of European protestant connections is strengthened by the strongly protestant preamble to Lott Stalling's will, atypical of Neat House gardeners' wills in general.[1]

The earliest gardens at the Neat Houses were on the Thames bank because this was the best location. Here was access to water transport, both to send the produce to market and to receive the large quantities of dung (which kept the soil fertile) from London streets and privies. The alluvial soil close to the river may also have been superior to that inland. The importance of Thames-side land was recognized in rents by 1675 when a piece of garden ground on the Thames bank might command a rent over £2 per acre more than land half a mile inland.[2]

The early records of the Gardeners' Company have a few Neat House men listed before 1620 and many more after. Examining the parish registers for entries concerning early gardeners known to be working at the Neat Houses before 1630 we find that many were neither born nor married in the parish and others can be traced back for only one generation. Only two surnames of early gardeners, Arnold and Weston occur frequently in the area before 1600 but the earliest Arnold identified as a gardener, in 1620, was born in Weybridge in Surrey and Weston was a very common surname in the parish.

1. Blanche Henrey, *British Botanical and Horticultural Literature before 1800*, Oxford, 1975, vol. I, pp. 157–8; C of WAC, F.345; Guildhall Library, MS 3389/1a; MS 21.128; GLRO, Arch Middx Orig. Wills 27 Feb 1684/5 Box 5.
2. C of WAC, 1049/10/Box 1/3

Later gardens were created piecemeal away from the riverside. John Weston took a lease in 1655 on three acres of established garden ground by the Thames and two acres of meadow north of this land which was to be 'laid to' it. A year later, his neighbour Thomas Bolton leased two more acres of meadow 'to be laid to the said three acres of garden ground'.

These are instances of established gardeners taking in new ground. However, new men also created gardens from former arable or pasture. Joshua and Richard Oldfield, youths from Derbyshire apprenticed to a Neat House gardener in 1633, stayed on after serving their time. Joshua spent considerable money and effort in creating a new garden for himself which he gradually expanded. In 1675 his 'new middle ground [was] brake upp but 8 years since'. Philip Luke, who died in 1682, son of a Hertfordshire yeoman, was probably originally an apprentice. Whatever the origins of the first few gardeners, former apprentices who made new gardens expanded the gardening area later in the century.[1]

1. C of WAC, 1049/10/1/3, 1049/3/18/11; Herts RO 5954, 5953; Guildhall Library, MS 3389/2; GLRO AM/PW/1682/33; Arch Midx Wills 20 Jan 1665/6.

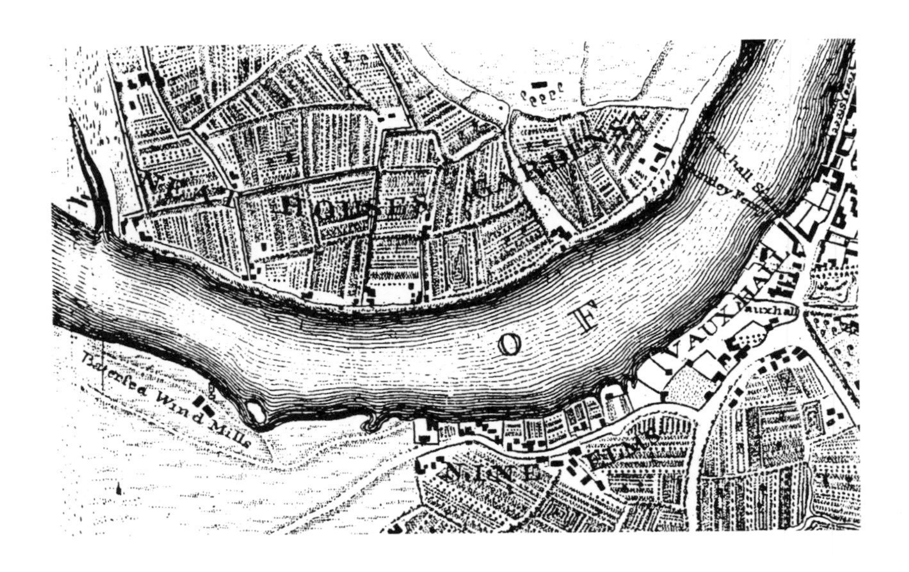

Figure 15. The Neat House gardens in the 1740s, from Rocque's *An Exact Survey of the Cities of London and Westminster*, 1741–6.

CHAPTER V

The gardens and their crops

'...wide rows of artichokes planted with Colley flower plants.'

All gardeners have to contend with the weather. The micro-climate of this part of Westminster was generally favourable to gardening. The Neat House gardens were 'situate on the south-west border of the town, consequently the cold of the north-west wind, so hurtful to vegetation in exposed situations, is considerably moderated, or rarefied, before it passes over these grounds: they are also naturally low and sheltered.' Although not close enough to the built-up area to benefit from the warmth created by big cities, the Neat Houses were at least downwind of most of the smoke generated by Westminster and London. The Chelsea Physic Garden today, only half a mile away and also by the Thames, contains species native to warmer climates growing outdoors all the year round.[1]

The second half of the seventeenth century witnessed a succession of bad winters. John Evelyn recorded 'very cold and severe weather' in the winter of 1683 and again in 1684 and 1685. The Thames was frozen each year, thick enough for a frost fair to be held on the ice. Evelyn wrote a letter to the Royal Society outlining the damage done to his garden at Deptford by the winter of 1683 and although he was concerned with shrubs and ornamental trees (and

1. John Middleton, *View of the Agriculture of Middlesex*, 1798, p. 262; P. Brimblecombe, *The Big Smoke*, 1989, p. 24.

his hibernating tortoise which was dug up 'stark dead'), damage was probably equally severe in commercial gardens.[1]

In the 1680s, some summers were as inclement as winters. After an 'excessive hot and dry spring,' the summer of 1684 was 'never so dry a season as in my remembrance,' the first significant rain not falling until 10th August. The next year saw a severe drought in spring and early summer, and the summer of 1686 was very wet. Several Neat House gardeners were in arrears with their rents in 1686; the bad weather over the previous three years may well have caused their financial distress. [2]

In the eighteenth century, Richard Bradley noticed the problems gardeners around London had with the weather. In May 1723:

> The Violence of the Weather at the Beginning of the Month was so severe, that I was inform'd by the Gardeners about the Neat-houses, that they could not cut one Fourth part of the Asparagus which they had done, even in the preceding month, which tho' bad enough, had afforded good Crops; the Hot-bed Cucumbers were in some places destroy'd, which gave the few that were brought to market as great a Price as when they first came in.

Bad weather made it difficult for the Neat House gardeners to pay the rent in the summer of 1745. The landlord's agent reported, 'it is a dismal Year among them, not having saved any Cucumbers or Melons'. Even so, the Neat House gardeners were some of the most skilled in London at counteracting both climate and weather: hence their reputation as growers of out-of-season vegetables.[3]

The Neat House gardens were a compact group, one garden bordering another. The main area was shaped like an inverted fan, with

1. G. Manley, 'Central England temperatures 1659–1973', *Quarterly Journal of the Royal Meteorological Society*, 100, 1974, pp. 389–405; *The diary of John Evelyn*, ed. E.S. de Beer, 1959, pp. 762–6, 785; John Evelyn, *A letter to the Royal Society*, reprinted Berkeley, USA, 1990, pp. 9–15.
2. *The diary of John Evelyn*, pp. 771, 775–6, 811, 839; C of WAC, 1049/3/5/69.
3. Richard Bradley, *General Treatise*, 1723, vol. I, p. 109; Chester CRO, Grosvenor Records, Box 77/2; Richard Bradley, *Philosophical Account*, p. 184.

the long curve of the Thames bank forming the southern perimeter, with a few just outside the boundaries—on the edge of Tothill Fields for instance. The surface of this area is covered by alluvium; in the words of a late-eighteenth century writer, 'This soil, no doubt, originally consisted of the sediment deposited there from the richly impregnated water of the Thames'. The ground is low-lying, and was, prior to gardening, a mixture of arable and pasture in the north, and wet meadow in the south. In 1675, the gardens were bounded by two small rivers, the Tatchbrook in the east and the Westbourne in the west. They were further dissected by minor streams and ditches which bore away surface water in the winter months. Adequate drainage was not achieved without effort: in 1675 some gardeners complained of the difficulties they experienced on low-lying land. James Tyndall proposed to 'raise' two of his northernmost acres by adding new soil, because they were under water and he faced 'great charges' in so doing. Drainage often required co-operation and forbearance: in the 1670s the landlord's agent noted, 'George Crockford must not be att any charge when his ground is broke up to mend ye water cource of Philip Luke & Edward Curtis which passes through Crockford ground whom is willing to bare ye Inconvenience of that: but not to be att any further charge by reason he is not a farthing ye better: but a great looser by reason of his ground'.[1]

In the summer, gardeners needed access to water. Two gardeners near Ebury Farm, some distance from the Thames, shared a pump to raise water and another was allowed a small encroachment on Lott Stalling's land, 'designed for a place to take upp water out of the said ditch'. Gardeners close to the river could irrigate their land with Thames water (figure 16): 'by a little attention to the sluices [they] fill their ditches, dipholes, and wells, with Thames water, and detain

1. C of WAC, 1049/12/114; 10/3/1691; W. Loftie Rutton, 'The Manor of Eia, or Eye next Westminster', *Archaeologia*, vol. LXII, 1910, p. 28; Charles T. Gatty, *Mary Davies and the Manor of Ebury*, 1921, vol. II, plate 31; *Geographers' A to Z Atlas of London and Suburbs*, 1990; O.S. 1" surface geology; Middleton, op. cit., p. 261; C of WAC, 1049/10/Box 1/3; 1049/6/1; 1049/10/Box 43.

Figure 16. The counterweighted bucket used by gardeners to raise water from their dip holes. Illustrated by John Middleton in his *View of the Agriculture of Middlesex*, 1798.

it in such places to within about eighteen inches of the surface, and by that means save a great deal of labour in the watering of their crops.' Stephen Switzer noted an additional advantage, 'River water, especially such as it is about London, or any great city, where it is continually disturb'd and made thick by its own motions, and the soil of the washings of the streets and grounds, is much better for watering than either spring or rain water.' Such irrigation was not always healthy. In 1623, some ascribed the outbreak of contagious spotted fever in London 'to the extraordinary quantity of cucumbers this year, which the gardeners, to hasten and bring forward, used to water out of the next ditches, which this dry time growing low, noisome and stinking, poisoned the fruit'.[1]

The Thames was tidal at the Neat Houses, kept at bay by an embankment which had been there since at least the fourteenth century. Because of the serious consequences of inundation, gardeners by the Thames were responsible for maintaining their section of the embankment and eviction was threatened if this duty was neglected. Along the bank were wharves and docks, depicted in the survey of

1. C of WAC 1049/3/8/22; 1049/3/8/26; 1049/3/20/14; Middleton, op. cit., p. 255; Stephen Switzer, *The Practical Kitchen Gardener*, 1727, p. 43; Charles Creighton, *History of Epidemics in Britain*, 1965, vol. II, p. 32.

1675 and mentioned in almost every gardener's lease. Gardeners fortunate enough to abut the Thames all had wharves; those without direct river access had wharfage rights written into their leases.[1]

Lanes and paths, frequently mentioned in leases, were vital to gardeners with no direct river access. The 1675 survey of the estate shows an apparently haphazard collection of minor lanes and paths, some of them dead ends, either branching off the main inland roads or leading from the Thames. When, however, one tallies these thoroughfares with individual holdings, it is clear they gave each gardener access to the Thames or a main road. As gardens were created, so lanes were made. Leases in 1670 and 1675 mention two new lanes, one giving access to the road to Westminster for the northern gardeners and a 'lane lately made and enlarged out of grounds of late occupied by Elizabeth Bell' which connected several holdings in the middle of the gardens with the river. The occupiers were responsible for the upkeep of their own boundaries and a variety of materials were employed, including stone or mud walls, hedges and paling fences.[2]

The acreage held by individual gardeners at the Neat Houses in 1675 varied from three acres to just over eleven. In 1677, two had just over 16 acres apiece; in 1700, one held 18 acres. Successful gardeners may from time to time have held larger amounts of land than their colleagues, but there was no tendency for holdings to increase at the expense of smaller enterprises. Each holding might contain several gardens. The 1675 survey depicts individual gardens varying in size from one to seven acres. They were subdivisions of the original arable and pasture fields and probably reflect the piece-meal conversion to horticulture in the course of the seventeenth century. The Neat House gardens in the late seventeenth and early eighteenth centuries thus formed a landscape of small, low-lying

1. Middleton, op. cit., p. 261; Loftie Rutton, op. cit., p. 39; C of WAC, 1049/5/1; 1049/3/5/66; 1049/10/Box 1/3.
2. C of WAC, 1049/12/114; 1049/3/23/7; 1049/3/18/3; 1049/10/Box 1/3; 1049/10/Box 43; 1049/3/18/35; GLRO MI.1718/10.

enclosures, intersected by ditches, streams and paths, with a scattering of houses, taverns, and garden sheds, softened by walls, hedges and a few trees.[1]

Most gardening books from the late sixteenth century onwards have an introductory section giving advice on choosing the site, laying out, and preparing the soil, of a kitchen garden. These books were written for gentlemen with private gardens, but the advice would have served for the intensive kitchen gardens at the Neat Houses. The initial task of any gardener was to make the soil fit for gardening. The whole area was dug over at least twice, once to remove weeds and again to add manure or other additives to create a good soil structure. Full trenching was usually advised: a deep trench was excavated at one side of the garden of at least two spades' blades depth and the soil beside it cast into the trench, creating another trench parallel to the first, and so on across the whole garden. This was a costly and time-consuming exercise but, 'tho these are chargeable ways of ordering of Gardens, yet it is done once for all, and the Charge will be abundantly answered in the Growth of what plants you set in it'.[2]

The internal layout of an efficient kitchen garden had to achieve a balance between using as much of the ground as possible for crops and providing sufficient access for the cultivation and harvesting of produce without trampling the soil or vegetables. By the sixteenth century, gentlemen's kitchen gardens were of a common pattern. Large paths to carry barrows and other large implements divided the garden into 'quarters', which were sometimes four areas separated by a cruciform path, but often six, eight or more oblong blocks. These in turn were divided into a number of beds or plots, separated by narrow paths a foot or so wide. Such arrangements can be seen

1. C of WAC, 1049/12/114; 1049/10/Box 43; 1049/7/2; 1049/10/32/1; 1049/12/114; Gatty, *ut sup*.
2. C of WAC, 1049/10/1/3; Gardiner, *Profitable Instructions*; J. Mortimer, *The Whole Art of Husbandry*, 1716, vol. II, pp. 112, 115; William Cobbett, *English Gardener*, 1829, paras. 22–25.

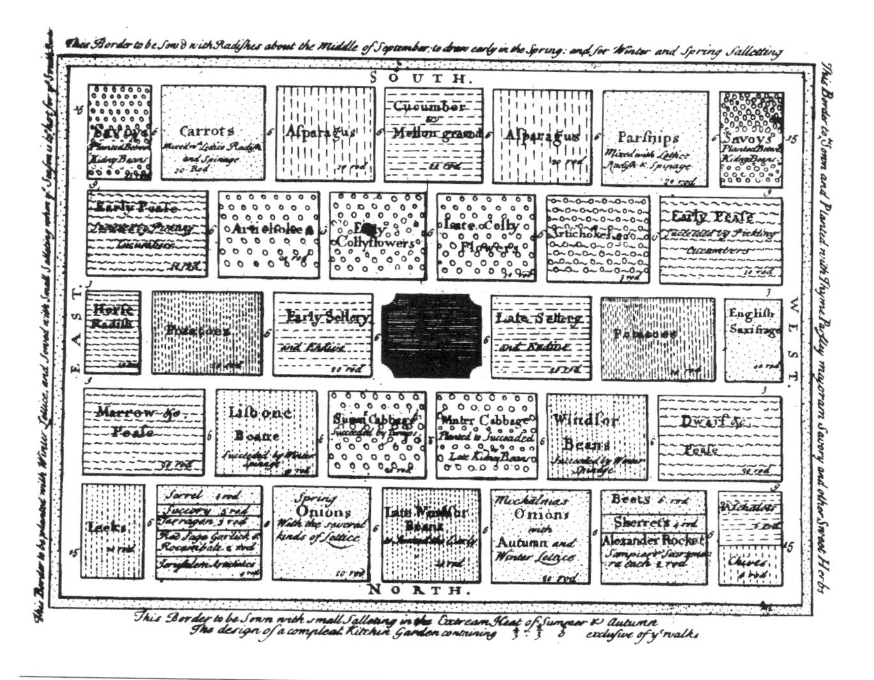

Figure 17. An ideal layout for a gentleman's large kitchen garden. Market gardens were worked as intensively as that depicted here. Illustration taken from Batty Langley, *New Principles of Gardening* (1728).

in many illustrations: William Lawson's ideal small garden of 1618; the frontispiece of Leonard Meager's *New Art Of Gardening* published in 1697; or in a particularly detailed plan of an ideal kitchen garden in Batty Langley's *New Principles of Gardening* of 1728 (figure 17). Beds were long and narrow. Richard Gardiner in 1598 specified a width of 45 inches and John Mortimer wrote that beds 'ought not to be wider than you can reach across', emphasizing the need to tend the crops without disturbing other plants.[1]

In Langley's plan, beds were between two and six rods in extent (an acre contains 160 rods, each of which is a square of sixteen feet

1. William Lawson, *The New Orchard and Garden*, 1618, p. 10; Mortimer, op. cit., p. 115; Gardiner, op. cit.; L. Fleming & A. Gore, *The English Garden*, 1979, pp. 66, 94.

and a half), and whole quarters of twenty or thirty rods contained larger vegetables such as potatoes, parsnips, carrots or cabbages. Meager also shows cabbages and cauliflowers in quarters or patches, whilst some beds are entirely covered with glass lights or cloches. Under the high wall around gentlemen's kitchen gardens were narrow beds planted to gain maximum benefit from shelter and the sun. The south-facing border of Langley's garden was to be sown with radishes in mid-September to catch the winter sun, whilst on the shady side the border was sown with small salleting in summer and autumn to avoid extreme heat.

* * * * * *

Maps of the period show commercial gardens arranged in beds and quarters as described by gardeners' manuals. Rocque's map of London, published in 1746 (figure 15), and his smaller scale map of London and ten miles around of 1754 are two good examples.[1] With the aid of the only really detailed probate inventory for a Neat House gardener yet discovered we can go inside a garden, test the veracity of Rocque's (and others') mapping conventions and confirm that commercial gardens were laid out like the kitchen gardens of the gentry. The surviving inventory details the goods and chattels of Robert Gascoine (figure 18).[2] On his death in 1718, he held two pieces of land from the Grosvenor estate and one from the Wise estate, all of it at the Neat Houses. He had no predecessors there and may have been a protestant refugee who fled France following the revocation of the Edict of Nantes in 1685.[3]

The two gardens rented from the Grosvenor estate are easily identified on the 1675 estate map as those formerly held by Edward

1. John Rocque, *An Exact Survey of the City's of London, Westminster, ye Borough of Southwark and the Country near 10 Miles Round London*, 1754; Ralph Hyde, *The A to Z of Georgian London*, 1981; D. Loggan, *View of Oxford*, 1675.
2. See appendix.
3. C of WAC, 1049/3/5/69; Chester CRO, Box 77/1/39; Loftie Rutton, op. cit., p. 48 fig 3; GLRO, AMPI 1718/10.

Curtis (see figure 14). The Home Ground, where Gascoine had his house, was a field of almost two and a half acres on the bank of the Thames. Its north-western corner just met his Middle Ground, a long field of approximately one and a half acres which was divided by an access path in 1675. This path had disappeared by 1718. His northernmost garden, part of the sub-manor of Neat and on the Wise estate, was located in the large triangular area opposite the site of the old Neat Manor House which in 1675 was owned by Robert Price. In the inventory, this is called 'the ground next to the Monster leading to Chelsea,' referring to the Chelsea road which ran along the top of the land, bending round the site of the old moat of the Neat House. The Monster was a tavern on the apex of the bend, across the road from the Neat House. It is shown on Horwood's map of 1799. This northerly garden was bordered to the north by the Willow Walk, the long, straight road along the top of the triangular piece of Wise land. The inventory also states that Gascoine's garden was next to a meadow owned by 'Mr Jones'. This meadow has to be the most westerly of the three fields comprising the Wise estate.[1]

William Jones, who produced Gascoine's inventory, was himself a gardener. His did his task methodically, first itemizing the contents of the house, then going round each of the three gardens, looking at the crops on the banks and in the borders and finally listing everything growing in the interiors. Following him round the Home Ground, we can discern its layout (figure 18).[2]

Jones found banks filled with cauliflower and cabbage plants along the western and northern sides of the garden. Such banks, probably earthed up against walls, caught the winter sun and were ideal for raising early seedlings. Further along the northern boundary the banks gave way to beds, probably narrow beds sheltered by the

1. GLRO AM/PI 1718/10; C of WAC, 11049/12/114; John Rocque, *A Plan of the Cities of London and Westminster*, 1746; Richard Horwood, *Plan of the Cities of London and Westminster*, 1799; Loftie Rutton, op. cit., pp. 48–9; GLRO AM/PI 1718/10; C of WAC, 11049/12/114; *Geographers' A to Z Atlas of London*, 1990.
2. C of WAC, 24/10/1700; GLRO AM/PI 1718/10.

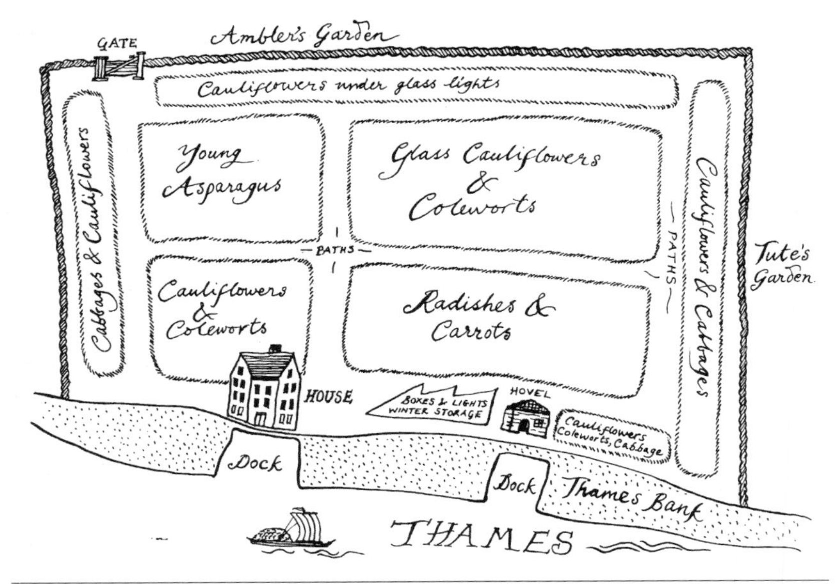

Figure 18. A reconstruction of the plan of Robert Gascoine's Home Ground, based on details given in the inventory taken on his death in 1718.

garden wall, again making the most of the winter sun. Banks occupied the north-eastern corner of the garden, with beds again towards the Thames. Few crops were growing by the Thames bank itself but much of it was occupied by the house and a wharf.[1]

Having perambulated the sides of the garden, doubtless following a path giving access to the beds and banks, Jones examined the interior. This was arranged in 'quarters'. Four were specifically mentioned by Jones, together with some beds of asparagus which may well have been in a fifth 'quarter'. Again, it is possible to follow him round as he moves from one quarter to another. At one end of the garden Jones found all the glassware not in use stored, probably in a shed or 'hovell'.

The other two gardens also had banks or borders around them. The interiors were made up of quarters, usually planted with one

1. GLRO AM/PI 1718/10; C of WAC, 1049/3/20.

crop only and ranging from 30 to 50 rods in area, 'pieces' which were of varying size (the smallest 18 rods, the largest 124 rods), and beds. The beds contained asparagus, much of it under glass.

The Middle Ground had a 'Middle dipping', which must have been a 'dip hole' like that described and illustrated by the agricultural reporter John Middleton in 1798, which he said was peculiar to market gardens in Middlesex. Dip holes were wells only a few feet deep because of the high water table. Water was raised by a bucket on a rope attached to the end of a pole which pivoted on a post and was counterweighted at the other end, allowing it to be raised with little effort. The same device in Egypt, a *shadduf*, raised water from the Nile for irrigation (figure 16).[1]

The northern garden ('the ground next to the Monster leading to Chelsea') was some distance from the house and outhouses of the Home Ground. Isolation necessitated covered storage space in the form of two 'hovells'. The garden also contained three 'read hedges and cross hedges to them athwart.' Middleton noted these portable hedges at the Neat Houses in 1798 (figure 19). Their use was well summarized by John Parkinson in 1629:

> And to prevent both the frostes, and the cold bitter windes which often spoyle their seede new sprung up, they use to set great high and large mattes made of reedes, tyed together, and fastened unto strong stakes, thrust into the ground to keepe them up from falling, or being blowne down with the winde; which large mattes they place on the North and East side to breake the force of these winds, and are so sure and safe defence, that a bricke wall cannot better defend anything under it, then this fence will.[2]

The disposition of beds, banks, and borders in Gascoine's market garden did, therefore, conform to the layouts advised in gardening manuals written for the gentry and to the gardens depicted on

1. Middleton, op. cit., p. 255; GLRO AM/PI 1718/10.
2. Middleton, op. cit., p. 259; Parkinson, op. cit., p. 463; GLRO AM/PI 1718/10; C of WAC, 1049/3/5/66.

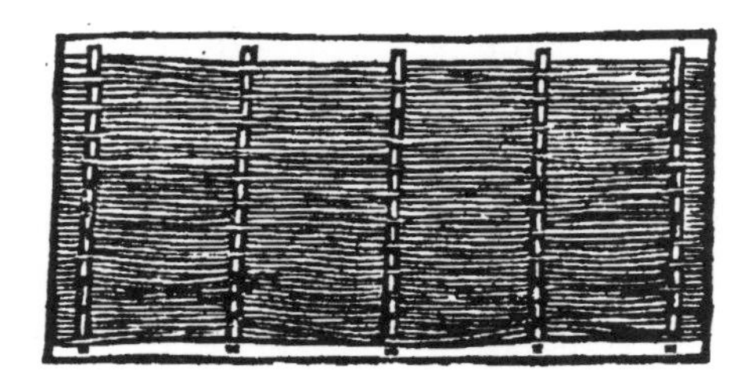

Figure 19. A straw mat similar to those used at the Neat Houses as wind breaks, as illustrated in *Le Jardinier Solitaire* (1706).

contemporary maps. The Middle Ground described above was said to occupy one acre, two rods and seven perches in a survey of 1723. It was probably all under crops when the probate inventory was taken in 1718 and quarters occupying 188 rods are listed. If we add 20 rods for a bank and border not measured, the total is 208 rods, 82 per cent of the total area given in 1723. Thus the growing area was roughly four-fifths of the garden, leaving one-fifth for paths and alleys. This may appear a high proportion of wasted space, but was a necessary sacrifice when constant access was required to all parts of the garden without damaging growing crops.[1]

Gascoine, and no doubt the other Neat House gardeners, practised row cultivation. Crops in rows allowed weeds to be eradicated by hoeing or hand pulling. Plants were given space to grow and could be thinned, earthed-up, and harvested without undue damage. Row cultivation also facilitated multiple cropping: sowing wide rows of one crop interspersed with another. The first crop matures and is harvested, which allows space for the second to grow. This is also interplanted with a third crop to follow on. Alternatively, fast maturing crops were planted in succession between slow growing ones. Several of the beds in Gascoine's garden in February 1718 were sown in this way: 'one quarter of glass colley flower plants adjoining to the high bank near Mr Tutes with some colwart between ym. & some young lettuce....One bank ag. the pale butting Mr Amblers

1. Loftie Rutton, op. cit., p. 48; GLRO AM/PI 1718/10.

house sowed with reddish & planted with colleyfflower plants...a bank next the Willow walk with wide rows of artichokes planted with Colley flower plants.' The technique was common in intensive London gardens at the time. On the other side of the Thames, Samuel James's market garden at Lambeth in October 1725 was full of intercropping: 'a parcel of Topps and cabbages on the same piece...a Bitt of Spinnage planted with Cabbage plants...Thirty one Rows of Sallary and Some Savoys...fifteen Rows of Salary and some later flowers...thirteen rows of Sallary & some Endiffe'.[1]

Intensive cultivation required careful soil management, to maintain fertility and prevent disease and pests. The 'great Garden-grounds in or near London' in 1670 had less problems with disease than most because 'their grounds are in a manner made new and fresh once in two or three years, by dung and soil, and good trenching.' John Strype observed in 1720 that the Neat House gardeners 'keeping the ground so rich by dunging it...make their crops very forward to their great profit'.[2]

Manure was the mainstay of the Neat Houses' productivity. The use of dung from the city's privies and streets as manure was already common before the first Neat House gardens were established. In 1593, one writer was concerned about the hazard to health posed by laystalls just outside the City. (Laystalls were the grounds where dung was left to mature.) As rehearsed above, the same author feared disease from produce 'sown upon those corrupt laystalls and grounds which many gardeners and others have of late practised to sow, before they have lain a convenient time to rot and be fit for manuring.' The Gardeners' Company, in a petition of 1617, was more positive, claiming they 'cleansed the City of all dung and noisomeness', although even in 1717 the fastidious complained that

1. Thomas Hill, *The Gardener's Labyrinth* (1577), ed. Richard Mabey, 1987, pp. 44, 80, 85; Richard Gardiner, *Profitable Instructions*; Joan Thirsk, *Agricultural Change*, p. 261; GLRO, AM/PI 1718/10; PRO, Prob 3/25/8.
2. John Strype, *A Survey of the Cities of London and Westminster*, Book VI, p. 78; Meager, *New Art Of Gardening*, 1697, pp. 164–5.

asparagus grown by London gardeners in beds of dung was 'of a Colour unnatural, and a Taste so strong and unsavoury'.[1]

The Thames greatly facilitated the movement of dung between City and gardens. By 1607, 'The soyle of the stables of London, especially neere the Thames side, is carried Westward by water, to Chelsey, Fulham, Battersey, Putney, and those parts for their sandie grounds.' In 1700, the association between boats on the Thames carrying dung and gardeners formed the basis of popular humour: the crew of a 'Western barge' shouted at a boatful of Lambeth gardeners, 'Foh, you nasty dogs that get your bread by the drippings of other people's fundaments; well you pray for the dunghill, for if that should fail you, no turd, no gardener. Who was that, you rogues, that dung'd his own cap at stocks-market, and carried home the old gold to enrich his radish-bed?'[2]

Fifty years later, better roads offered another route. The observant Swedish botanist Pehr Kalm noted in March 1748 that refuse from the London streets was, 'shovelled together in heaps, and laid in the dung-wagons to be carried out of town to some particular place where they are shot....When farmers and others convey anything into the town to be sold, they seldom drive with an empty load home, but they mostly take a wagon full of this manure out with them from the places where it is collected together.' Laystalls, where dung was stored and matured, were sometimes specifically mentioned in Neat House leases.[3]

Perhaps reflecting the increasing use of manure in the gardens, the Grosvenor estate in September 1685 asked for legal opinion on the ownership of dung which lay in the beds and cucumber holes of a garden when a tenancy ended. Counsel's opinion was that the dung

1. BL, Lansdowne MS 74, ff. 75–6, my thanks to Joan Thirsk for this reference; W.T. Crosweller, *The Gardeners' Company: A Short Chronological History, 1605–1907*, 1908, p. 11; John Laurence, *Gentleman's Recreation*, 1717, p. 90.
2. John Norden, *The Surveyor's Dialogue*, 1607, p. 191; Frank Muir, *Oxford Book of Humorous Prose*, Oxford, 1990, pp. 29–30.
3. Pehr Kalm, *Kalm's Account of his Visit to England on his way to America in 1748*, ed. J. Lucas, 1892, pp. 143–4.

in 'the Cucumber Hole may bee lawfully taken away by him that layd itt there, But as for the dung in ye Ridges itt ought to goe with ye land.' Despite this opinion, a gardener on the estate a century later was allowed to remove dung from the land at the end of his lease, undertaking that 'wherever the Mould or manure is taken away he will level the Ground or fill up the Pitts.'[1]

Apart from use as a general fertilizer, the major use of dung at the Neat Houses, especially stable litter, was for hotbeds. Indeed, the dung was often used for hotbeds first and then as a soil additive. A hotbed is simply a layer of dung covered with soil. The dung ferments, producing heat which warms the soil and accelerates the growth of seeds and plants. The addition of glass to cover the growing plants increases efficiency. In the words of Richard Bradley in 1724, 'A Hot Bed is the common Help made use of by Gardeners to forward the Growth of a plant, and force Vegetation, when the Season of itself is not warm enough. By this means (if it be managed with Skill) the hottest Climate of the World may be so nearly imitated, that seeds of Plants brought from any Country between the Tropics may be made to flourish.' The technique was used by the Romans and beds were made in England from at least the sixteenth century. Their commercial use in kitchen gardens, however, appears to date from the mid-seventeenth century, when glassware was cheap enough to be viable in market gardens.[2]

Hotbeds were at first made by placing the dung in a trench. As time went on, trenches became shallower and beds were increasingly formed from piles of dung on the surface, avoiding compaction and increasing the flow of oxygen to the dung, which aided fermentation and increased the heat. The dung was topped by a layer of fine soil, ideally a 'light fresh loam' the depth varying with the crops to be raised and the heat required. The heat usually lasted for about two

1. C of WAC, 1049/3/5/73; 1049/5/1.
2. Philip Miller, *Gardeners Dictionary*, 1763; Richard Bradley, *New Improvements of Planting and Gardening*, 1724, pp. 373–4; David Stuart, *The Kitchen Garden*, 1987, p. 39.

months, but varied according to the season and the dung used. The life of hotbeds could be extended: 'when it begins to cool, cut away the Sides of it slope-ways, and laying fresh Dung to them, it recovers its Lost Heat, which is call'd Backing or Lining of a Bed; and this Work skilful Gardeners will repeat some-times five or six times in a Season, as they see occasion, rather than make a fresh bed.' A variant was the cucumber hole, a device for raising early cucumbers. In March, 'make holes about the bigness of a bushel or bigger, the which you are to fill up with warm stable-dung, setting it close, making a hole in the midst in which you are to Plant three or four Cucumber Plants with their mould about them.'[1]

* * * * * *

Cucumbers in 'holes' were forced under glass. Hotbeds of all kinds required glass. Bell glasses (also called hand or moving glasses), covered cucumbers in their holes. These bell-shaped covers, about two feet in diameter at the bottom, with a knob of glass at the top for handling, were ideal for covering one or two plants. Large hotbeds were usually covered by frames: 'The common hot-bed frame is a rectangular box of wood, bottomless and highest at the side to be placed to the north, subdivided by crossbars dovetailed into the outer frame, and each subdivision covered by a glazed sash.' The 'glazed sashes' were flat frames with panes of glass, often called lights, set in them, the whole assemblage being known to gardeners as 'boxes and lights'. These could cover a greater area than glass bells and a corner of the frame could be raised off the boxes by wedges to let in air on mild days.[2]

Patents to establish an English glass industry were issued in 1552 but all glass, including the green glass used for bell glasses and garden lights was expensive for some time thereafter. Sixteenth-

1. Miller, op. cit.; Stuart, op. cit., p. 41; Bradley, op. cit., pp. 373–9; Meager, op. cit., p. 199.
2. J.C. Loudon, *Encyclopaedia of Gardening*, 1824, pp. 299–303.

century books, including the very practical treatise on vegetable production by Richard Gardiner of 1599, do not mention the use of glass. The gardening section of *The English Husbandman* of 1635 describes sowing cucumbers on hotbeds, but recommends only that they be covered at night with 'a Mat, Canvasse, or other covering, which being placed upon stakes over the dung bed shall every night after Sunne-set be spread over the same.' This procedure limits sowing to springtime, when the days were already quite warm.[1]

Advice given by Leonard Meager in the 1670s implies that glass was still expensive but within the pocket of a modest gentleman (or prosperous London gardener). Bell glasses are advocated for melons, covered with a penthouse of poles and mats, and also for cucumbers, 'but if you cannot afford to Glass them, you must not Plant out until the weather be very warm and dry.' Stephen Blake, in *The Complete Gardeners Practice* of 1664, a book with many references to techniques used in London commercial gardens, has similar advice.

The amounts in use were small. Henry Rosemary, a Bermondsey gardener, left glass worth £6 10s when he died in 1673; John Harvest of Chelsea had 150 [bell] glasses worth £4 on his death in 1681. At the Neat Houses, probate inventories of gardeners also included moderate valuations for glassware: William Pearce, January 1679, £18; Philip Luke, February 1684, £16; and John Lee, July 1684, £3 2s. Kathleen Weston, whose garden was taken over as a going concern for non-payment of rent in 1687, had the following: '7 doz Old Glasses, One Dozen of Bell Glasses & two ould Glasse fframes,' valued at £2 10s.[2]

English glass production expanded towards the end of the century. An estimated 240,000 dozen green glass bottles were produced in

1. Joan Thirsk, *Economic policy and projects*, 1978, p. 34; Richard Gardiner, *Profitable Instructions*; Gervaise Markham, *The Second Booke of the English Husbandman*, 1635, pp. 18–19.
2. Meager, op. cit., pp. 193, 199; GLRO, AM/PI(1)1681/11; AM/PI (1)1682/18; AM/PI (1) 1682/14 ; AM/PI 1688/85; C of WAC, 1049/3/5/66.

Figure 20. A bell glass with removable straw covering, as illustrated in *Le Jardinier Solitaire* (1706).

1695 and were 'sold for half the prices they were a few years since [because]...the glass maker hath found a quicker way of making it out of things which cost him little or nothing.' Until the nineteenth century, London was the centre of glassmaking, and London kitchen gardeners benefited from easy access to this fragile product.[1]

The increased availability of glass is reflected in the amount held by Robert Gascoine on his death in February 1718. No valuations were made but Gascoine had boxes and lights stored for the winter, some others in use in all three of his gardens, together with '1240 whole bellglasses', in all three grounds. There are no later inventories from the Neat Houses, but frequent references to vegetables forced under glass by these gardeners testify to their widespread use of hotbeds and glass in the eighteenth century.[2]

Evidence of the importance of glass to kitchen gardeners comes from south of the Thames. A freak hailstorm in July 1750 caused havoc on the Surrey bank, from Lambeth, opposite the Neat Houses, down to Southwark and inland to Camberwell and Newington. Forty three gardeners' losses were quantified in inventories. In three-quarters, the glassware accounted for much of the damage. The total

1. John Cary, *An Essay on the State of England*, 1695, quoted in J. Thirsk & J.P. Cooper, *Seventeenth Century Economic Documents*, 1972, p. 321; Francis Buckley, *Old English Glasshouses*, 1915, p. 19.
2. GLRO AM/PI 1718/10.

damage to glassware was over £2,800, averaging more than £67 per gardener. Eleven individuals lost more than £100 of glass each. With bell glasses priced at 3s each and glass lights at 7s a frame, these figures represent a considerable investment.[1]

* * * * * *

By these techniques, kitchen gardeners lengthened the growing seasons of many vegetables, supplying a burgeoning market among the upper and middle classes. Richard Bradley mentions in 1719 hot-beds and glass used for growing cucumbers, melons, early carrots, asparagus, winter peas, winter beans, and small mixed sallad stuff. He had 'heard that a gardener about Westminster has reciev'd above thirty Pounds in one Week for forc'd Asparagus', and commented 'that the Pride of the Gardeners about London chiefly consist in the Production of Melons and Cucumbers at times either before or after the Natural Season'. Cauliflowers were raised under glass with much success: 'This excellent plant is cultivated in our Kitchen-Gardens with so much Skill, as to have them for the use of the Kitchen above half the Year'.[2]

The Neat House gardeners were important innovators in this respect. Bradley mentioned one, a Mr Jewell, who sent cauliflowers 'of the fine sort to market first, about the 14th [of May 1723]'. This same man, to stimulate demand with something new, 'was the first Gardener in England that raised the young Sallad Herbs for the Winter Markets, and Kidney-Beans in Hot-beds'. In 1721, Bradley found that they 'abound in Salads, early Cucumbers, Colliflowers, Melons, Winter Asparagus, and almost every Herb fitting the Table'. If there was profit, innovation quickly followed. When white cabbage was more prized than green, they copied some provincial private gardeners who folded the leaves of 'Coleworts or strong

1. Surrey County Record Office, Quarter Sessions Transactions, Midsummer 1750.
2. Bradley, *New Improvements*, pp. 108, 116, 127, 138, 156, 160; *Adam's Luxury & Eve's Cookery*, 1744, pp. 57, 64.

Cabbage Plants' and tied them together, 'by which Means, in a Fortnight's Time, the inner Parts will become white, and eat as well as any Cabbage'. By 1721, 'most of the gardeners about the Neat Houses are fallen into that Method, and have reap'd good Sums of Money from it.'[1]

Enhanced income and reputation for novelty was matched by the gains from breaking the straitjacket of season. Even in the dead of the winter of 1718, Robert Gascoine's garden contained growing crops of cauliflowers, radishes, carrots, coleworts, lettuces, Michaelmas onions, corn salad, spinage, artichokes, and celery, as well as asparagus.

Not only table vegetables were grown. The Neat House gardeners experimented with unusual crops, some of which one would have thought unsuited to their low-lying land with its high water table. In the 1660s there were said to be 'acres' of liquorice growing at 'the Neat Houses nigh London'. This crop needed skill, labour and time to bring to maturity, as well as careful handling between harvesting and sale. Only the one reference has been so far discovered, and the crop did not last long in the gardeners' repertoire, but it is further evidence of innovative spirit. At that juncture, several writers were advocating liquorice culture. Either profits from liquorice were particularly high, or gardeners were influenced by writers' recommendations.[2]

Innovation was combined with the long experience and collective wisdom of a community of gardeners who had worked for many generations on the same site. This depth of knowledge was recognized by Bradley who thought, 'there is no where so good a School for a Kitchen Gardener as this Place'. The nobility sent apprentices

1. Bradley, *General Treatise*, 1723, vol. I, pp. 109; 1726 edition, vol. I, p. 157; John Strype, *A Survey of the Cities of London and Westminster*, Book VI, p. 78; Bradley, *Philosophical Account*, p. 184.
2. Bradley, *Philosophical Account*, p. 184; GLRO, AM/PI 1718/10; Stephen Blake, *The Complete Gardeners Practice,* 1664, pp. 106–7; PRO, Prob 4, 2316; Prob 4, 2260; Walter Blith, *The English Improver Improved*, 1653, p. 351.

from their own gardens to the Neat Houses to learn the skills needed to produce good crops. One such apprentice wrote a book on kitchen gardening, edited and published by Stephen Switzer in 1727.[1]

The Neat Houses were especially renowned in the seventeenth and early eighteenth centuries for their melons and asparagus. They grew musk melons, the variety best suited to English weather, although still difficult to produce—needing heat, hand pollination, careful ripening and constant protection from pests and diseases. As Thomas Hill complained in 1577, 'The Mellons and Pompions hardly come up in this Country, at due time of the yeare, without labour, cost and diligence of the Gardener in hastening them forward.' John Gerard, writing in 1597, observed that, 'They delight in hot regions'. He had not raised them himself, merely seen ripened musk melons grown by Mr Fowle, keeper of the Queen's house in St James's.[2]

An Italian immigrant, Giacomo Castelvetro, claimed in 1614 that the melon was his favourite fruit, 'for no other reason than its marvellous sweet scent, the most wonderful perfume in the world'. Eighty-five years later Evelyn was similarly smitten by the 'Paragon with the noblest productions of the Garden'. The gentry liked melons. Such demand, from those who could afford to pay for luxury, encouraged gardeners to grow them. Initial results were disappointing: the Venetian Ambassador reported in September 1618 that he had visited 'a sorry melon ground' near London where he paid 35 pence each for melons. John Parkinson, however, detected an increase in production as growing techniques were mastered in the 1620s: 'This Country hath not had untill of late yeares the skill to nourse them up kindly, but now there are many that are so well experienced therein, and have their ground so well prepared, as that they will not misse any yeare, if it be not too extreme unkindly.' John Evelyn, 79 years old in 1699, remembered, 'that this fruit was

1. Bradley, ibid.; Switzer, *The Practical Kitchen Gardener*, p. vii.
2. GLRO, AM/PI 1718/10; Bradley, ibid.; David Stuart, op. cit., pp. 182–4; Hill, ed. Mabey, op. cit., pp. 192–5; John Gerard, op. cit., p. 918.

very rarely cultivated in England, so as to bring it to Maturity,...I myself remembering, when an ordinary Melon would have been sold for five or six Shillings.' Commercial gardeners were most likely responsible for the enlarged production and discovering how to grow melons successfully. Hotbeds formed of rotting dung were important ingredients, although some mid-seventeenth-century writers disliked the 'unwholesom use of Mucke-beds here in London'.[1]

Gerard's mention of melons in 1597 ties in with an incident early in the reign of James I. At a banquet given to celebrate peace with Spain in August 1604, the Spanish envoy was proudly presented with a ripe, English-grown melon, which was divided between him, the King and the Queen. This fruit probably came from the same royal garden known to Gerard. Given the proximity of the royal court to the Neat Houses, and the obvious interest there in melons, might not the Neat Houses have received stimulus from both a demand for melons amongst courtiers and contact with their gardeners who possessed the skills to grow them?[2]

Neat House gardens were certainly early producers of melons, and were renowned for the fruit by 1632. Samuel Pepys bought one there in 1666 and they were still regarded as their speciality in the 1720s and 1740s. They were in many ways ideally suited to melon production: large and regular supplies of dung by boat from London, and the glasses used in winter for other crops could cover the melons in the summer. There were labourers to tend the melons and, more importantly, the gardeners and their experienced journeymen had the skill to raise them. Melons were an attraction to the gentry who

1. Giacomo Castelvetro, *The Fruit, Herbs & Vegetables of Italy*, trans. Gillian Riley, 1989 p. 88; John Evelyn, *Acetaria*, 1699, p. 37; *State Papers Venetian*, 1617–19, 1909, p. 319; Parkinson, op. cit., p. 525. By 1824 melons were 5s to 20s each in June at Covent Garden, falling to 3s in September and October (J.C. Loudon, *Encyclopaedia of Gardening*, 1824, p. 1063); Robert Sharrock, *The History of the Propagation of Vegetables*, Oxford, 1660, p. 131; A. Speed, *Adam out of Eden*, 1659, p. 97.
2. W.B. Rye, *England as seen by Foreigners*, 1865, pp. 119–121.

visited the gardens for relaxation or went to the pleasure gardens and taverns nearby.[1]

Asparagus, like melons, needed much dung and labour. Gerard refers in 1599 to cultivated asparagus as 'manured or garden Sperge'. In 1629, Parkinson thought asparagus in 'much esteeme with all sorts of persons'. It retained its status as a 'superior vegetable' into the following century, despite large-scale production around London.[2]

Early gardening books describe the cultivation of maincrop, summer asparagus produced on many acres at Battersea, Deptford and other suburban areas which, 'coming so early, and continuing so long,' gave a steady supply to the markets. John Worlidge, in 1675, mentions an innovation, 'If you take up the old Roots of Asparagus about the beginning of January, and plant them on a Hot Bed, and well defend them from Frosts, you may have Asparagus at Candlemas [February 2nd], which is yearly experimented by some.' In 1682, Leonard Meager found asparagus being forced in this way by London growers. These writers may have had the Neat Houses in mind. Three February inventories, in 1682, 1687 and 1718, indicate considerable investment in raising winter asparagus.[3]

One gardener had £32 worth of asparagus in his garden. Another, Mrs Weston, was growing 156 rods of asparagus (almost an acre) valued at £31 4s: 80 per cent of the total value of her growing crops. It was four times as valuable per acre as the 54 rods of spinach and onions which formed her next most significant crops. Asparagus also figures prominently in Robert Gascoine's inventory of 1718: 196 rods were growing in seven different parts of the garden. This is at least one and a quarter acres in a garden which may only have been

1. Philip Massinger, *The City Madam*, 1658, Act III, Scene i; *The Diary of Samuel Pepys,* ed. H.B. Wheatley, vol. V, p. 365; Bradley, ibid.; Chester RO, Grosvenor Records Box 77/2.
2. Gerard, op. cit., p. 1111; Parkinson, op. cit., p. 503; J.C. Loudon, op. cit., pp. 643–4; Maisie Brown, *The Market Gardens of Barnes and Mortlake*, 1985, p. 14.
3. John Worlidge, *Systema Agriculturae*, 1675, pp. 151–2; Loudon, op. cit., p. 647.

about eight acres in extent: a high investment of dung, bell glasses, and labour in a single crop. Not all was being forced for the winter market. There were also beds of young asparagus, and some 'of the first years planting', probably covered with dung and soil to protect it from frost, needing at least two more years before maturity. For winter cropping, he had 32 rods of 'cutting sparrow grass' and three beds of 'forc'ed Sparrow grass under boxes and lights'.[1]

Winter asparagus was raised from three- or four-year old roots transplanted on to hotbeds. Used for a season, the roots were exhausted. Neat House gardeners always had 'great store' of these fledgling roots which 'they sell to one another, when any one of their fraternity wants them, for about four or five shillings per pole'.[2]

1. GLRO, AM/PI (1) 1682/14; AM/PI 1718/10; C of WAC, 1049/3/5/66.
2. Switzer, *Practical Kitchen Gardener*, p. 177.

CHAPTER VI
Labourers and apprentices

'Honest and laborious Gardeners'

All kitchen gardening was labour intensive, but the Neat House gardeners probably hired more men and women per acre than most of their London competitors. No specific records of employment survive and assumptions have been made using several historical sources. These include the Middlesex wage assessments for gardeners of 1666; a draft rating assessment of 1664 which lists servants and apprentices living-in at the Neat Houses; the known acreages of each gardener's holding in the late seventeenth century; a comment on the wages of London garden labourers in 1747; and John Middleton's reports of estimates of labour costs per acre at the Neat Houses made in 1798.[1]

The Middlesex wage assessments provide the official rates of pay for the gardeners' adult male workforce: those employed by the year were to be paid £6, £4, or £3, according to skill, and day labourers either 2s 6d or 1s 6d without food or 2s or 1s with food, again according to skill. These were official and therefore minimum wages. In 1747 gardeners' journeymen were said to be paid between 9s and 15s a week, presumably for day labour without food.[2]

1. Middleton, op. cit., pp. 264, 380–2; L. Martindale, 'Demography and Land use in Middlesex in the late Seventeenth and Eighteenth Centuries in Middlesex', PhD, London, 1968, Appendix XI, pp. 460–63; R. Campbell, *The London Tradesman*, 1747, p. 274; C of WAC, F4538.
2. Middleton, ibid.; Martindale, ibid. Although the Neat Houses were in Westminster, Middlesex was only a short distance to the west, and it is safe to assume a similarity between Middlesex and Neat Houses wages.

Although these official figures cover a broad band of rates, they can act as basis for a plausible model of employment in the first twenty years of the reign of Charles II. Annual rates of pay were much less than day rates because they were paid to journeymen who lived-in. The day rates without food must provide a more reliable guide to labour costs.

To obtain the figures in table 3, I have multiplied the acreage known to have been rented by each gardener from the Mary Davies estate in 1675 by an estimate of labour costs. These were £35 per acre, according to John Middleton in 1798. Allowing for inflation over the intervening years 1675–1798, this could be restated as £20 per acre in 1675.

The total hypothetical labour costs of Lott Stalling, with a holding of 6.5 acres, are therefore 6.5 x £20: £130. His colleague William Weston, with a holding of six acres, would have had to pay an estimated 6 x £20: £120.

We can divide these totals by the wages of a second-class gardener, as laid down by the Middlesex justices in 1664, and thus achieve an estimate of the number of men employed. Such a man was paid at least 1s 6d per day, and may be expected to work 300 days every year. His annual wage, therefore, is £22 10s.

Lott Stalling's total labour cost of £130, divided by an individual wage of £22 10s, gives a figure of 5.78 employees. William Weston's costs of £120 result in a figure of 5.33 employees.

This does not yet allow us to calculate the total number of people working on each holding. We can use the figures so far obtained to suggest an overall average of 0.89 full-time employed males per acre. But this does not take into account the unpaid labour of family and apprentices. If we add the known adult family of each gardener, his female servants, and an estimated one apprentice per holding, we arrive at the figures in column 3 of the table. Thus a large garden like that of John Weston might have more than twelve people working on it the year round.

Table 3.[1] Estimated labour engaged in the Neat House gardens, *c.* 1660–80.

Gardener	*Rent divided by day labour rate*	*Wife & adult children & female servants plus est. 1 apprentice*	*Total labour per garden*
Edward Boyton	2.6	5	7.6
Edward Curtis	4	2	6
William Tute	4	1	5
William Grey	6.2	5	11.2
John Packer	2.7	5	7.7
Lott Stalling	5.8	2	7.8
William Weston	5.4	3	8.4
John Weston	9.3	3	12.3

Such figures, however, hide sizeable fluctuations. The permanent workforce usually lived-in but in the summer months many women, old men and children were taken on at day or piece rates in London gardens; the Gardeners' Company estimated that its members employed many thousands of such people in 1617. In 1798, John Middleton noted the same practice, especially for such unskilled seasonal work as weeding and carrying produce to market, activities which employed a great many Welsh women in the summer at a shilling a day. Despite the insecurity of piecework, he thought the life of day-labourers near London very pleasant. More meat was consumed in Middlesex than in England generally, and vegetables were plentiful. He estimated per capita consumption of beer at a staggering 100 gallons per annum. In his eyes, labourers were less amenable to discipline here than elsewhere, because, 'gentlemen's servants are mostly a bad set, and the great number of them kept in this country, is a means of the rural labourers acquiring a degree of idleness and insolence unknown in places remote from the

1. Middleton, op. cit., p. 380; Martindale, ibid.; C of WAC, F4538; 1049/10/Box 1/3.

metropolis.' Labour, skilled and unskilled, full-time and seasonal, was thus a formidable cost for the Neat House gardeners; in 1798 it was the largest single item of expenditure, an estimated 44.75 per cent of total annual working costs.[1]

London gardeners' apprentices were generally taken on at a premium of between £5 and £10 in the mid-eighteenth century. Several Neat House gardeners' sons were apprenticed to their own fathers, but some went to other Neat House masters. Many apprentices came from outside the area: George and Joshua Oldfield, sons of a Derbyshire husbandman, were apprenticed in 1633 and eventually became masters in their own right at the Neat Houses. Others mentioned in the 1630s came from Saffron Walden in Essex, Skevington (?Skeffington in Leicestershire), Aldenham in Hertfordshire, and Kildare in Ireland, so the Neat Houses were drawing apprentices from the country before Bradley recommended the education available.[2]

1. W.T. Crosweller, *The Gardeners' Company: A Short Chronological History, 1605–1907*, 1908, p. 11; Middleton, op. cit., pp. 256, 264, 382, 387–90.
2. Campbell, op. cit., p. 335; Bradley, *Philosophical Account*, p. 184; Guildhall Library, MS 3389.1A.

CHAPTER VII
Rents, tithes and taxes

'...lettyth the same ground to them that will geve most mony ffor e'ry Acre.'

Location, fertility, market access, demand from alternative uses, landlord and tenant improvements, and institutional rigidities or flexibilities all affected the level of rents at the Neat Houses, before and after the arrival of market gardening. The parish of St Martin-in-the-Fields, first as a suburb of Westminster and then as a part of greater London, saw its land go through more changes of use than most parishes in England. As more profitable uses of its land supplanted existing ones, rents rose.

In 1485 the best arable in the area let for 8d to 1s an acre and meadow for 2s an acre. By 1549, arable was 2s 8d to 3s 4d and meadow up to 8s an acre. Allowing for inflation, this is almost double the earlier rent. Meadow rents rose most in this comparison, and parishioners in 1549 complained of the decay of tillage: demand from the capital for meat and dairy produce was the cause. The petition also protested at rack-renting on the manor of Neat by the Crown's tenant, Sir Anthony Browne, who 'lettyth the same ground to them that will geve most mony ffor e'ry Acre'. The petitioners recognized that, following the confiscation of Westminster Abbey lands, the piecemeal distribution of land by the Crown to courtiers had destroyed any customary constraints on rent increases and restrictions on change of land use. This freedom was exploited by new landlords who let to tenants willing and able to pay higher rents for land which they could use to produce the food demanded by expanding London markets. Such institutional conditions were ideal for the emergence of market gardening and it is significant that the first gardens appeared on land in the manor of Neat mentioned by the petitioners.[1]

The first identifiable gardener was paying £3 an acre for his land in 1622, when arable in the area let for £1 and meadow at between £2 and more than £3 per acre. Wet meadow by the Thames commanded the highest rent and it was this land which first attracted the attention of the gardeners because of its fertility and favourable location. Garden and wet meadow rents were similar until the mid-seventeenth century; thereafter they increasingly diverged as prices for meat and animal products rose less and the demand for good quality garden produce expanded. In the 1670s, wet meadow yielded £2 to just over £3 per acre whereas garden ground produced £6–8.[2]

A rent review in 1675 provides fascinating details of the negotiations between the landlord and tenant gardeners which preceded the issue of new leases. The estate was administered by John Tregonwell, guardian of Mary Davies, Thomas Grosvenor's future bride. The gardeners occupied most of the southern half of the estate which, although Mary Davies held the freehold, was let on a long lease at no annual rent. The lease expired in 1675, at a time when the Davies estate was in need of ready cash, and the administrators found themselves for the first time able to extract sizeable rents from the gardening tenants.[3]

Tregonwell instructed his agent to discuss new terms with the gardeners, their own leases having expired at the same time as the head lease. The agent had, initially, no clear idea of the gardeners' circumstances. He drew up a rough set of proposals for new rents,

1. John McMaster, *A short history of the Royal parish of St Martin-in-the-Fields*,, 1916, pp. 254–9; Peter J. Bowden, *Economic change: wages, profits and rents, 1500–1750,* Cambridge, 1990, p. 150.
2. C of WAC, 1049/1/28; 1049/10/Box 1/3; 1049/10/Box 43.
3. Gatty, op. cit., vol. II, p. 190.

expressed in round sums of between £5 and £8 per acre with the areas of each garden rounded to the nearest half acre. Divided into three broad categories—land by the waterside, land remote from the waterside, and land more remote from the waterside—the proposals recognized the premium on land with good river access.[1]

In terms of annual rent the proposed increases were steep; most were between 33.3 per cent and 100 per cent higher than before. The old leases, however, included premiums or entry fines. John Weston paid an entry fine of £24 and a rent of £20 for 16 years from 1655; John Lee's old rent was £12, but he had to pay a fine of £9 for a seven-year lease in 1668. The new leases were to be for only 3 years, much shorter than the old ones, with no entry fines, although the agent used the term 'fine' in comparing the old and proposed new rents. Thus Lott Stalling's old rent per acre was approximately £4 15s, and the proposed rent would be £7 per acre. The old rent for his holding was £31 10s per annum, and the agent sought to impose a notional fine of one year's old rent, £31 10s. This fine was to be paid over two and a half years, that is £12 3s 4d per annum. This sum, added to the old rent, gave the proposed new rent of £43 3s 4d. The agent made similar calculations for all the other gardeners, although the size of the 'fines' based on the old rents varied and the calculations were also worked out in terms of rents per acre. In performing these curious calculations the agent may have been trying to compare the combination of fine and rent in the old leases with the rent-only provisions of the new leases, both for his own benefit and to show the gardeners that the overall effect of the proposed increases were not as steep as they first appeared. John Lee's new rent was to be £16. This was equivalent to his old rent of £12, plus a £10 fine over two and a half years, in comparison with his old fine of £9 for a seven-year lease.[2]

1. C of WAC, 1049/10/Box 1/3; 1049/12/114.
2. C of WAC, 1049/10/Box 1/3.

Why the increases were divided by two and a half years is a puzzle; the terms of the subsequent leases are for three years and do not specify any abatement for the final six months for unexhausted improvements and cultivations. Although no entry fines are included in the new leases, the estate did try to extract some money in advance from the gardeners for the new rents. In the July preceding the Michaelmas when the new leases were issued, the gardeners were asked for advances. John Ambler paid £24 out of the proposed new rent of £44 10s, and James Tyndall parted with £24, a year's old rent. Philip Luke refused 'to give any advance, so told him must provide agt. Michalmass'. These advance payments may be, in effect, entry fines which the estate did not wish formally to embody in the leases. The estate's need for money at the time may account for the apparently *ad hoc* arrangements. The shorter, three-year leases were in line with current practice.[1]

Having formulated his proposals, the agent put them to the gardeners, at the office of a lawyer, Mr Mosier, on 27 July 1675. All gardeners on the estate bar four were present and many argued vigorously against the new rents. The agent was at some disadvantage; he was meeting the gardeners for the first time (he had to call a roll and tick off their names to identify them) whereas they were part of an established, close-knit community. Some tenants argued that the acreages on which the new rents of their holdings were based were too large, and the agent noted, 'all tenants aggrieved in Measure to have their owne Surveyor & ours to measure againe'.

The agent recognized in his initial proposals the importance of river access but the gardeners introduced him to the complexity of this aspect of rental value. Robert Chandler complained that he had no land by the waterside, his nearest land was '150 paces from the Water'. He wanted improved access to his northernmost holding, either via a new bridge over the Meadow Ditch, or by the right to use John Packer's cartway and then to cut through James Oldfield's

1. C of WAC, 1049/10/Box 1/3; Bowden, op. cit., p. 262; Gatty, op. cit., vol. I, p. 178.

garden to get to his main garden and thence to the Thames. Such arrangements required delicate negotiations. The agent noted that Oldfield was to be charged no rent for 'a way for Chandler through his ground'. The paths and tracks so important for access were maintained by the gardeners themselves but they derived varying benefits from them and wished the rents to reflect this. John Weston, it was noted, 'joynes to repair wayes but only hath the use of wayes to one of his three leases'. In the end the agent noted that the new leases must contain covenants specifying responsibility for repairing the 'ways' as well as the ditches which drained the land.[1]

The agent had not considered soil conditions in his initial proposals. Edward Curtis refused to accept his new rent because part of his land was sandy. James Oldfield's middle ground was 'so Cold by lying much under water that 20£ at a tyme he hath lost'. Robert Chandler and James Tyndall also claimed their land was waterlogged. Some gardeners complained they had spent much on improvements to their holdings and that the rents should be abated accordingly. John Weston had 'broke up' two acres of meadow land himself to convert it to garden and James Oldfield had done the same. James Oldfield's father 'spent himself & all he had to make ye ground'. Even so, his house was 'ready to fall downe' and others sought consideration for onerous repairs which they had to carry out under the terms of a lease: John Ambler and Edward Curtis both had to repair part of the Thames embankment. John Tyndall was to raise two acres to combat wetness, and build a new house and wharf. William Tute agreed to his new rent 'wth some Complain of a hard bargaine,' Lott Stalling 'would have had 20s abate but agreed,' and most of the gardeners came to some agreement at that meeting, or were content to have their grievances considered. Edward Curtis, however, was obdurate. Asked to give £31 10s per annum 'he cryes out that that is Exaction' and 'will not give ye rent demd'.[2]

1. C of WAC, 1049/10/Box 1/3; 1049/12/114.
2. C of WAC, 1049/10/Box 1/3.

Following the meeting with the gardeners, the agent consulted his master and final proposals were put to another gathering of the tenants in January 1675/6. Most gardeners secured some reduction on the original rents proposed. Several received abatements of £1 or £2 because of cold or wet ground. John Weston, who negotiated also for his son William, obtained £2 off William's rent and £10 10s off his own bill as recognition of work he had done on his land. As he and his son together were to pay £120 per annum, his bargaining power was strong. Chandler obtained access through Oldfield's land, together with a rent reduction of £8 5s 8d. Oldfield was excused payment for the quarter-acre of land covered by Chandler's new track and the general inconvenience of the arrangement. John Lee saved £2 'in consideration of his house to bee by him well repaired,' and Alexander Gray agreed to pay £3 10s less than originally demanded for making a new fence on one side of his land.[1]

Two gardeners who fought strongly against the new rents profited from their obstinacy. John Ambler was at first asked to pay £44 10s per annum. A reduction to £39 was proposed. On his refusal to agree at the second meeting, he was threatened with ejection. The landlord, however, eventually accepted £35. Of Edward Curtis it was decided, 'That he bee ejected, if hee not comply to hold his land at £30 per Annum.' He accepted in the end, but he saved £5: £1 per acre on the original proposal.[2]

Both sides gained something: rents were raised, but almost all the gardeners argued some reduction on the initial proposals. Some gained significant concessions. Complaints were heeded, and many were accommodated in the new leases. Although these were technically 'full repairing', abatements were granted for onerous house repairs and for unexhausted improvements to the land. The gardeners undertook long-term investment in raising low-lying land to combat flooding, showing confidence in their security of tenure and,

1. C of WAC, 1049/10/Box 43.
2. C of WAC, ibid.

despite apparent vulnerability when their leases fell in, they were in a strong position. Social cohesion and trading experience gave them mutual confidence in their dealings with the new landlord.

Despite the small acreages involved, the rents under discussion were not small and the gardeners themselves were, by the standards of the time, substantial businessmen. The landlord could threaten to eject the tenant if rents were not agreed but this potent weapon was two-edged. The intensively cropped gardens had a delicate balance of irrigation and drainage, and yielded the profits to pay high rents only with massive applications of dung and much expenditure on labour to weed and dig, tend hotbeds and dung heaps, plant, sow, harvest, and market crops. Continuity of cultivation was essential and Sir Thomas Grosvenor was as anxious to 'secure a tenant who would buy the stock and rent the land from Michaelmas' as he was to obtain the rent owed by a gardener he was ejecting in 1686. Just a few weeks without a tenant could, as the steward of the estate recognized in 1745, lead a garden to 'irretrievably run to ruin'.[1]

A similar process of negotiation in 1677 preceded the new leases the following year, resulting in the same rents for some gardeners although most paid a pound or two less for their land. Thereafter, rents fell. In 1700, gardeners on the Grosvenor estate paid between £4 4s and £5 12s an acre; reductions of one quarter to one third on the rents of 1675. The figures reflect lower profits because of a run of bad weather. With the fall in rents came a return to longer leases. The 1678 leases for garden ground were granted for seven years. Some seven-year leases continued to be agreed into the first decade of the eighteenth century, but a 21 year lease was granted to a gardener in 1684, another in 1687, and most of those found from the first three decades of the eighteenth century are for 21 years, fixing rents for longer periods.[2]

1. Bowden, pp. 250–273; C of WAC, 1049/3/5/69, Chester CRO, Box 77/2.
2. C of WAC, 1049/10/Box 43; 1049/3/5/66; 1049/3/5/88; 1049/3/5/69; 1049/3/30/47; 1049/3/30/41; 1049/3/30/12; 1049/3/30/30; 1049/3/31/5; 1049/3/18/22; 1049/3/18/26; 1049/3/19/3; 1049/3/20/14; 1049/3/20/20; Chester CRO, Box 77/1.

The gardeners were, from the point of view of the Grosvenors, and no doubt the other landlords at the Neat Houses, a considerable source of income during this period. The total annual rent due from the gardeners at Michaelmas 1675 was £631 14s; at Lady Day 1677, it was £569; and in May 1679, £475 16s. The total due from the whole of the Grosvenors' Westminster lands in 1679 was £1357 4s 9d. The gardeners, occupying only a small part of the estate, accounted for 35 per cent of the rental.[1]

The landlord often had to wait for his rent from tenants because, despite the intensity with which the gardeners worked the land, most vegetables were produced and sold in the summer. Rent due at Christmas was often paid in the following summer, along with the rent then due. Thereafter arrears again built up. Sir Thomas Grosvenor advised his steward at the end of July 1686, ' Be brisk with them now which is the time for the gardeners to pay for a months hence they well pay none til Spring.'[2]

Many gardeners were in difficulties in 1686–7 when two years of exceptionally cold, dry winters and hot, dry summers—followed by an outbreak of fever in London, which affected many well-off customers—hindered both the production of and demand for Neat House vegetables. Nine gardeners were listed as in arrears in March 1684, for sums ranging from £24 to £121. Of these, five were still in difficulties in 1686 and some businesses failed.[3]

Katherine Weston's five-acre holding was taken in hand by Sir Thomas Grosvenor in February 1687. Her husband died in 1684 and she took over 10 acres of garden ground, held on two leases: the largest holding in the estate. Either she found the business too large to cope with, or the death of her husband left her in financial difficulties, for she had already disposed of the lease of five acres to Margaret Pretty for £100 by December 1684.[4]

1. Chester CRO, Box 77/1; C of WAC, 1049/10/Box 1/3.
2. C of WAC, 1049/6/1; 1049/3/5/69.
3. C of WAC, 1049/10/Box 43; 1049/3/5/69.
4. GLRO, AM/PI(1),1684/26; C of WAC, 1049/3/5/66; 1049/3/18/14.

William Crockford had money problems in July 1686 when another gardener was being sought to take over his land. He surrendered his lease in September owing £40, having been arrested for debt by Grosvenor. The terms of the surrender allowed him to retain his house and a 'boarded shed or drinking place'—one of the many beer houses in the area—at a rent of £4. The stock on the garden was to be appraised and any money gained from its disposal went first to pay off arrears, the surplus accruing to Crockford. One has the impression that, in difficult times, Crockford had used the rent money to keep the business going and eventually the landlord lost patience. Edward Curtis, a third failure, was a difficult tenant for whom Grosvenor had little sympathy. His case is discussed below.[1]

The high gross output of the Neat House gardeners should have made tithes a bigger annual expense than rent; one tenth of the estimated market value of the annual produce in 1798 would have averaged £22 per acre. In practice, the burden was far less. Most of the garden ground was contained within the old manor of Ebury or the sub-manor of Neat. Research occasioned by an attempt to increase vicarial tithes in the early nineteenth century found that landholders in the manor of Ebury had paid no tithe at all since at least 1537. The vicar of the parish surrendered the tithes to the monks at Westminster in 1314 and the lands became tithe free on the surrender of the manor to the Crown at the time of the dissolution. The Neat House tithes were also surrendered in 1314, but the Grosvenor tenants here paid a composition or modus in lieu of tithes averaging 4s an acre. In 1759 gardeners on the sub-manor of the Neat paid a total of £19 in tithe. In 1803, £10 5s was paid on 75 acres. In 1803 a valuer estimated potential tithes from the gardeners, if legal action succeeded, of £53 from the Neat and £37 from Ebury, a total of £90. This sum is modest, between 12s and 15s an acre. It must have been based on a tenth of the rent, not a tithe on the gross produce. Was the absence of tithes or a low composition an

1. C of WAC, 1049/3/5/69; 1049/3/5/68.

additional attraction to the gardeners who first started in the area in the seventeenth century? At Colchester in the seventeenth century, gardeners occupying land within the precincts of the castle claimed to be exempt from tithe because they were within the jurisdiction of the defunct castle chapel. The high gross output of gardeners made tithe-free land very attractive.[1]

Land tax, introduced in 1692 and payable by the tenant on the gross rent of the land, fell more heavily upon gardeners. The rate in the eighteenth century varied between 1s and 4s in the pound. A receipt for the tax in the Grosvenor records dated April 1736 records tax at 2s in the pound, with gardeners paying a total of between £1 4s and £3 each.[2]

1. Middleton, op. cit., p. 264; C of WAC, 1049/10/32/1; PRO, E 143,3 Wm & Mary, Mich.5, Essex.
2. Chester CRO, Box 77/2; W.R. Ward, *The English Land Tax in the Eighteenth Century*, 1953, p. 7.

CHAPTER VIII
Supplementary income

'...hurry me to the Burse, or Old Exchange
The Neathouses for musk-melons, and the gardens
Where we traffic for asparagus...'

It may seem strange that the only known side-occupation of the Neat House gardeners was opening their gardens to the public and providing alcohol, food and general entertainment, but they were only making money from the delight which Londoners took in having fun in the open air. Citizens enjoyed bowling alleys, archery butts, bear baiting, and open-air theatres in the sixteenth century and took more simple recreation 'at gardens and in the fields'. The next century saw the development of specialist pleasure gardens, such as the Mulberry Garden in Westminster (originally the base for James I's silkworm project) and those at Vauxhall, where food and drink, tobacco, music (and prostitutes), were available. Pleasure gardens were even more popular in the eighteenth century when both large and small were opened to satisfy public demand.[1]

Massinger's play *The City Madam*, published in 1632 but written earlier, contains the first reference to the Neat Houses as places of resort. They were then a fashionable destination for well-to-do Londoners and retained their reputation into the 1660s when Samuel Pepys visited them several times. The Grosvenor estate archives and

1. E. B. Chancellor, *The Pleasure Haunts of London*, 1925, p. 193 ; *The Diary of Samuel Pepys*, ed. H. B. Wheatley, 1946, vol. IV, p. 118, vol. VIII, pp. 22, 30, 380; M. Dorothy George, *London Life in the Eighteenth Century*, 1966, pp. 279, 296 ; BL, Lansdowne MS. 74, ff. 75–6.

Figure 21. The Monster Tea Gardens in 1851. They are reached through the arch on the right of the tavern. This was probably a pleasure garden in 1718, when it is mentioned as being next to Robert Gascoine's Middle Ground. The water-colour is by Thomas Hosmer Shepherd.

two probate inventories from the 1680s prove that the market gardens themselves, as well as taverns with gardens nearby, were pleasure gardens at that time. Later evidence is scarce: they may have lost their custom to the larger and more organized pleasure gardens at Vauxhall and Ranelagh or to the numerous other tea gardens, wells, and other open-air establishments which appeared in more convenient parts of the suburbs.[1]

The few references to pleasure gardens and taverns at the Neat Houses show them either along the main road from Westminster to Chelsea on the northern edge of the gardening area, or to the south

1. J. Massinger, *The City Madam*, pp. ix, 39; *The Diary of Samuel Pepys*, *ut. sup.*, vol. V, p. 365, vol. VIII, p. 30, vol. VII, p. 51; GLRO, AM/PI (1) 1684/26; AM/PI (1) 1687/13; AM/PI 1718/10; C of WAC, 1049/3/5/68; 1049/3/30/22; 1049/31/19; 1049/31/20; H. Hobhouse, *Thomas Cubitt: Master Builder*, 1971, p. 169; Dorothy George, ibid.; E. B. Chancellor, op. cit., p. 193.

along the Thames bank. The well-to-do of London and Westminster could reach the first by coach and the second by water. There seems to have been a division in the character of the establishments along these geographical lines. Those in the north were clustered around the site of the manor house of Neat near Ebury Bridge. By the time of a survey in 1723, a small hamlet had grown up there and it is possible the old manor gardens themselves may have been a place of resort. Jenny's Whim, a tavern by Ebury Bridge, was fashionable in the time of George II (1727-60). Its main attraction was a garden with a bowling green, cockpit, ducking pond, and discrete alcoves within which were, 'mechanical figures which, on a spring being touched, started up, to the terror and confusion of many'. This tavern may have been on the site of an earlier one called The Monster, which was next to Robert Gascoine's garden in 1718. None of the references to these northern places of pleasure connect them with working market gardens and none were known to have been run by gardeners. Samuel Pepys probably went to one of these taverns by coach in August 1667 after a play, with his wife and a leading actress in tow, 'there, in a box in a tree, we sat and talked and eat'.[1]

The southern pleasure gardens were more modest affairs, apparently run by some of the Neat House gardeners as extensions of their businesses. In essence, some of the gardeners opened their gates to the public and provided refreshments and places to sit. Those near the Thames bank were ideally situated to benefit from this additional source of income. Three gardeners, James Oldfield, John Weston, and William Crockford, were engaged in the business in the 1680s. All had gardens with a Thames frontage, and tenements on the bank from which to dispense drink. The map drawn as part of the estate survey of 1675 shows a curious, square building with a high-pitched, pointed roof like a small kiosk on John Weston's land. It was perhaps a summer house for his guests. Beer was the only

1. *The Diary of Samuel Pepys*, *ut sup*., vol. VII, p. 51; Chancellor, ibid., p. 368; Horwood Map, 1799; GLRO, AM/PI 1718/10.

beverage dispensed by Weston, Oldfield and Crockford. The latter sold his drinks from 'a boarded shed or drinking place' according to the record of 1686. Weston's stock of beer, 14 barrels worth £8 8s, some 784 pints, could have coped with quite a few thirsty customers, but there is nothing else in his inventory to indicate more than a modest bar-room. Crockford's 'drinking shed', together with his house and another tenement, was worth £4 a year when a rent was negotiated in 1684, which was less than the rent of one acre of garden ground.

The Neat Houses were clearly in fashion in the 1620s when the wench in Massinger's play longed for a coach:

> To hurry me to the Burse, or Old Exchange,
> The Neathouses for musk-melons, and the gardens
> Where we traffic for asparagus,...

but Pepys thought them somewhat tame when, in May 1668, he went with business companions,

> first to one of the Neat-houses, where walked in the garden, but nothing but a bottle of wine to be had, though pleased with seeing the garden, and so to Fox Hall [Vauxhall] where with pleasure we walked...there sat and sang, and brought great many gallants and fine people about us....We did by and by eat and drink what we had, and very merry.

The larger, more organized pleasure gardens such as Vauxhall held more attractions to those whose main interest was music, wine, and good company rather than well laid-out vegetable gardens.[1]

1. GLRO, AM/PI (1) 1684/26; AM/PI (1) 1687/13: C of WAC, 1049/3/5/88; 1049/12/114; *The Diary of Samuel Pepys*, *ut sup.*, vol. VIII, p. 30; Massinger, ibid.

CHAPTER IX
Wealth, living standards and society

'...a poor, painful widow...'

The wealth of the Neat House gardeners in the late seventeenth and early eighteenth centuries is difficult to gauge because of the lack of records. Some information can be gained from the eight probate inventories and seven wills that have survived, together with the 1664 hearth tax assessment and several rating assessments and details of the size of their houses and other small pieces of information which have come to light.

The total valuations of the inventories are not particularly high, with the exception of Robert Gascoine's in 1718. They are misleading, however, because only Gascoine's and John Lee's appraisers listed crops in full. Philip Luke's inventory mentions only his asparagus; in the others, there are no crops. (By law, most garden crops were not to be listed in a probate inventory.) When, three years after his death, John Weston's widow was forced to vacate the holding because of rent arrears, her crops, implements and seeds were valued at £56 18s 6d by her landlord, such a sum should be added to his inventory to gain fairer impression of his wealth.[1]

1. The inventory totals are: John Lee, 1684, £20 7s; John Weston, 1684, £31 11s 6d; James Oldfield, 1687, £20; Philip Luke, 1682, £124 11s 1d; William Chandler, 1688, £7; Lott Stalling, 1685, £38 13s; William Pearce, 1688, £18 3s 2d; Robert Gascoine, 1718, £198 18s 10d; GLRO, AM/PI (1)1684/26; AM/PI (1) 1687/13; AM/PI (1) 1682/14; AM/PI (1) 1688/20; AM/PI (1) 1685/15; AM/PI (1) 1688/85; AM/PI 1718 /10; AM/PI (1) 1684/93; M. Spufford, 'The limitations of the probate inventory', in *English Rural Society*, eds. John Chartres & David Hey, Cambridge, 1990, pp. 139–74.

Probate inventories do not list real estate. Wills do, but by no means all such property had to be included in a will. A gardener with a modest inventory total might have considerable property. In our case, only one inventory is paired with a will, that of Philip Luke who died in 1682. Luke bequeathed £106 in cash, compared with his total inventory valuation of £124 11s 1d. He also left his sister a messuage standing in one acre called Cripps Place in Aldenham, Hertfordshire, his niece a two-acre copyhold close in the same village, and his wife the lease of a house and lands at Borehamwood. Long leases of these and other lands were acquired by Luke in 1672 and 1674 for a total of £158.

No inventory survives for the gardener William Arnold, who died at the Neat Houses in 1644, but his will shows that he too had invested in land. In addition to cash bequests of £284, he left a lease of land in the area with an annual rent of £67 3s 4d, 'a lease from divers Inhabitants of Wandsworth of lands there which are let at £20 per annum,' and the reversion of the lease of his own garden ground and that of George Bayley. Whilst not amounting to the country estate to which successful middle-class Londoners aspired, Luke's property makes him much wealthier than his inventory would suggest, and if he and Arnold were typical of the Neat House gardeners, we must suppose others had property to provide a source of rent and a store of wealth. We should, however, also note that inventories do not list debts. Gardening was risky, and we have already noticed some casualties of bad weather who failed to pay their rent.[1]

Houses are another guide to wealth. In 1675, William Morgan surveyed the estate of Mary Davies and produced a map on which he drew small elevations of the principal houses, including those at the Neat Houses (figure 22). The survey, faded and in some places illegible, is nevertheless an important guide. Although the pictures

1. Spufford, op. cit., pp. 142, 151; Peter Earle, *The Making of the English Middle Class,* 1989, pp. 154–5; GLRO, AM/PI (1) 1682/14; Guildhall Library, MS 3389/2; Hertfordshire RO, 5953, 5954; C of WAC, 1049/3/5/68; 1049/3/5/66.

Table 4. Housing: information from the 1664 hearth tax assessment, a rate assessment of 1664, pictures of houses on the 1675 survey, and rooms listed in wills and inventories.

	1664 hearth tax: hearths[a]	*1664 rate assessment: adults in household [b]*	*1675 estate survey: storeys[c]*	*Wills & inventories: rooms[d]*
Edward Boyden	5	6		
Lott Stalling	7	5		6
William Pare	6	2		
William Apsley	4	3		
John Parker	6	5		
John Bayley	6	3		
Widow Crockford	3	1	1	
Edward Bell	6	7		
John Thare	6	6		
Philip Luke	4		2	*c.* 3
Richard White	2	4	2	
Widow Grey	9	11		
Edward Curtis	3	4	3	*c.* 7
William Tute	3	2		
James Tyndall	4		2	
Richard Oldfield	8			
Joshua Oldfield	4	4	2	2
George Crockford	2	3	2	
William Weston		5	2	
John Hall		2		
John Weston		5	3	3
John Lee		3	1	2
Christian Hall		5		
Widow Hall			1	

[a] Those known to be gardeners from Grosvenor estate records or other sources.
[b] As [a], see table 5 for analysis of status of adults listed.
[c] Counted from the pictures of houses on the survey (figure 22), garrets are counted as a storey.
[d] The tally includes rooms listed specifically in a will or inventory. Obvious rooms that have been omitted are added [i.e. if a bedroom is listed but no hall].

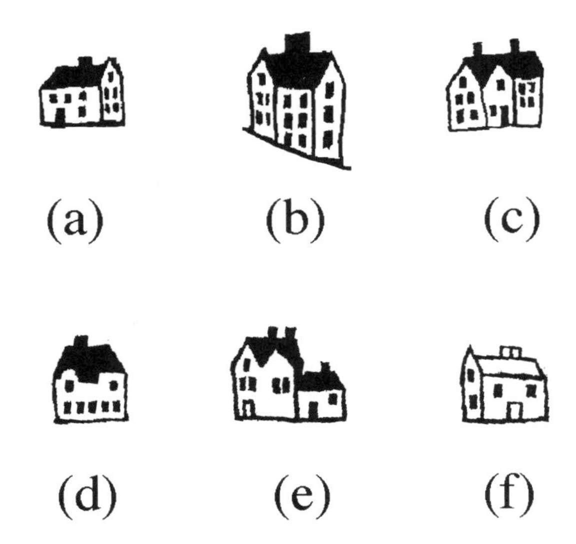

Figure 22. The gardeners' houses, redrawn from the 1675 survey of the Mary Davies estate. (a) George Crockford's house; (b) Edward Curtis's house; (c) Richard Oldfield's house; (d) William Weston's house; (e) Philip Luke's house; (f) Richard White's house.

are small, they were accurately drawn and tie in with impressions gained from the rooms listed in gardeners' probate inventories.[1]

The houses drawn on the 1675 survey were less than seventy years old, built by the first gardeners who carved out their holdings from meadow and arable land. The modest nature of John Lee's one-storey house is emphasized by his exemption from hearth tax in 1664 and the fact that he had no servants or apprentices living with him. Mrs Hall also had a small house with a small strip of land attached, but she had probably retired from active gardening and was described in an estate document as a 'poor painful widow'. Other gardeners lived in two- or three-storey houses, often with two chimney stacks, and many had garrets. Lott Stalling's house is not discernible on the survey, but he was assessed at seven hearths in 1664,

1. C of WAC, 1049/12/114.

and his inventory in 1684 listed goods in a three-storey building with two garrets, two or three chambers, a kitchen and a dining-room—the use of this word, rather than hall or parlour, indicates a degree of social sophistication. The inventory of Robert Gascoine, who died in 1718 in Edward Curtis' former house, suggests at least seven rooms, maybe more, in this large, three-storey house by the river.

The number of adults listed in rating assessments as residents could only have been accommodated in sizeable dwellings, although several servants, of the same sex, commonly occupied a single bedroom. The gardeners built and repaired their houses at their own expense, although the state of the buildings was taken into account when rents were negotiated. Some of the houses were in disrepair in 1675: James Oldfield claimed his house was 'ready to fall down' and James Tyndall proposed to build a new house if the landlord abated his rent. In 1677, John Weston was offered a substantial rent reduction if he undertook to build a brick house. This suggests that the first houses were timber-framed with wattle and daub, showing their age by the 1670s.[1]

Apart from a few omissions—Widow Hall and John Lee because they were too poor, the brothers Weston probably by evasion—the 1664 hearth tax provides a good overview. Of eighteen assessed, ten lived in houses with four, five or six hearths; five had two or three hearths; one each had seven, eight and nine hearths. Seventeenth-century houses had significantly more rooms than assessed hearths so that four to six hearths implies at least six to eight rooms. In terms of rooms, the average Neat House gardener and his family, that is his wife, children and servants, was at least as well housed as a Kentish yeoman.[2]

The analysis of the 1664 draft rating assessment in table 5 reveals that whilst some households were small, eleven out of eighteen houses were home to four or more adults, plus an unknown number

1. C of WAC, 1049/10/Box 1/3; 1049/10/Box 43; 1049/12/114.
2. Joan Thirsk, *Agricultural Change*, *ut sup.*, pp. 145–6; GLRO, MR/TH/7.

Table 5. Household composition: adults included in the 1664 draft rating assessment.

	Wife	*Children*	*Servants*	*Lodgers*	*Total [a]*
Lott Stalling	1		3		5
John Bayley	1		1		3
Edward Boyton	1		4		6
John Thare	1	2	2		6
Widdow Crockford					1
George Crockford	1			1[b]	3
Edward Curtis	1		2		4
John Hall	1				2
John Weston	1		3		5
WilliamTute				1[c]	2
Joshua Oldfield	1	1	1		4
Richard White	1	2			4
William Grey	1	3	2	4[d]	11
John Parker	1	2	1		5
William Apsley	1			1	3
Christian Hall		3		1	5
Edward Bell		1	5		7
John Lee	1	1			3

Notes

[a] I have added the gardener, head of the family but not mentioned in the assessment, to the total. Only gardeners identified from other sources included.

[b] A relative, Hugh Crockford.

[c] Mother of William Tute.

[d] Grey's lodgers were 'The Lady Scot: her husband lives in Kent,' and her 3 servants.

of children under sixteen years. If we allow the proportion of juveniles to be 25 per cent of the adult population, this results in an average household of 5.5 persons. The structure of the gardeners' families was conventional: a nuclear family of husband and wife with children before marriage living together. Only William Tute had an elderly parent living with him and he was unmarried at the time. Other widows, such as Mary Crockford and Widow Hall, lived apart from their married children.[1]

Goods listed in probate inventories provide a glimpse of conditions inside the houses. Mary Crockford, widow of William, one of the early gardeners, had few possessions on her death in 1666, sharing the only goods of value, some clothing and bedding, four pewter dishes and her fire irons, between her children. John Lee's household goods consisted mostly of bedding and modest furniture when he died in 1684, valued altogether at £5 11s. The goods of John Weston and Philip Luke, who both died in the 1680s, convey an impression of reasonable living. Pewter and brass were used for cooking and dining, there were enough tables and chairs downstairs, and at least the main bed had a good stock of bedding and curtains to keep out the draught. But some gardeners were prosperous enough to buy the domestic goods which were appearing in London shops and increasingly were found in the houses of the urban middle classes. Lott Stalling's pewterware weighed two hundredweight and his kitchen was well equipped with pots and pans. Brass candlesticks lit his dining room, where pictures and a looking-glass adorned the walls and carpets covered the floors. Fourteen chairs surrounded what must have been a substantial dining-table, served from a side-table with a court cupboard to store tableware.

Robert Gascoine died later than the other gardeners whose inventories survive. He possessed many more goods which contributed to domestic comfort, reflecting the growth of trade in consumer goods in eighteenth-century London. Despite damage to the inventory

1. Cof WAC, F4538.

which has obliterated much of the text relating to the chambers and parlour, we know he kept in these rooms caned chairs, half a dozen silver-hafted knives, a looking-glass and a clock. The kitchen goods reflect modest prosperity: amongst the usual array of brass, pewter and iron, were items to show that Gascoine and his family enjoyed new middle-class luxuries—a glass punchbowl and jug, a brass tea-kettle on a stand over a spirit lamp, and a copper coffee-pot. Gascoine also possessed some books: many gardeners of an earlier generation were unable to sign their wills.[1]

We have no intimate records, diaries or the like, to inform us of the relationship between the gardeners and their children, or indeed with the other members of their households. The apprenticeship of some sons to fathers or to other Neat House gardeners, and the occurrence of some colleagues' daughters as servants in gardeners' households at least indicates that adolescent children were content to remain close to their parents. In their wills, the gardeners made provision for all children. John Thare, leaving only 12d to each of his children in 1667, adds that he has already advanced 'their respective portions' to them during his lifetime lest he should be considered mean spirited. The poor widow Mary Crockford shared possessions amongst her children and directed her son George to have her garden and lease valued and 'among my six children equally divided'. Richard Oldfield instructed his wife to pay his only son £150 when he was twenty-one. He also left Richard a more personal bequest of family treasures which betrays some affection, 'one twenty shillings peece of Gold one ten Shillings peece of Gold one five shillings peece of Silver two Silver Spoones and one Silver box with about ten shillings of Small money in it'.

Only one instance of bastardy has been found. In 1644, William Arnold acknowledged a natural daughter who was still a baby. She

1. GLRO, AM/PW/1666/14; AM/PI (1) 1685/15; AM/PI (1) 684/26; AM/PI 1682/14; AM/PI 1685/15; AM/PI 1718/10; AM/PI 1684/93.

was well provided with leases and rents together with a mortgage and, after the death of his wife, she was to inherit half his land.[1]

Even less information on the relationship between husband and wife in gardening households is available. William Arnold left his wife the leases of his working garden ground despite his illegitimate daughter. Other men left the bulk of their estates, including the working garden, to their wives, and several wives continued in business after the death of their husbands, as London custom entitled them to do. James Oldfield, in a verbal declaration when close to death, made his wife executrix, 'she being a dear loving wife to me', which sounds like an informal expression of affection. It is likely that wives would be involved in the day-to-day running of the gardens and a close working relationship is to be expected.[2]

Servants lived in, as they did in the households both of middle-class London tradesmen and of the more substantial husbandmen and yeomen in the countryside. Just under a quarter of the adults in gardeners' households were servants. This is a smaller proportion than in the average middle-class Londoner's household. Our sample of gardeners, however, includes a widow living on her own and two or three poor individuals with no servants. In any case, our sample is very small.[3]

About half the resident servants in 1664 were male, either apprentices (some the sons of neighbours) or unmarried journeymen. The female servants would have been employed on domestic tasks but no doubt undertook garden work in the summer. The historian Peter Earle found that domestic servants in this period were in some demand and were said to do less work than in earlier times. He concluded that, on the whole, the relationship between servants and masters in London was good and there is no reason to believe this was not also the case at the Neat Houses. William Arnold desired his

1. Guildhall Library, MS 3389/2; GLRO, AM/PW/1670/122; AM/PW/1666/14; Arch Middx Wills 2 Jan 1665/6.
2. Guildhall Library, MS 3389/2; GLRO, AM/PW/1687/41; Earle, op. cit., p. 160.
3. Earle, op. cit., pp. 214–15; C of WAC, F4538.

four servants to carry his body to the grave and gave them £1 apiece. Other gardeners made similar bequests, usually £1, but Philip Luke left £5 to a former apprentice and a similar amount to that servant's brother who was still an apprentice.[1]

Lodgers were quite common in London houses at this time. Five of the eighteen gardeners' households had them, a similar proportion to that found by Peter Earle in London as a whole. In two cases, the lodgers were relatives. Lodgers, particularly if they were upper class, might contribute a significant amount to family income. William Grey had a genteel lodger in his large house in 1664: 'The Lady Scot', whose husband lived in Kent, together with her three servants.[2]

The gardeners and their households had close links with each other. Many were kith and kin. The Oldfield brothers, Joshua and Richard, originally apprentices from Derbyshire, were both in business in the 1660s, as were the widow Mary Crockford and her son George. In 1675, father and son John and William Weston were each gardening on separate holdings, the father negotiating a new rent on behalf of his less experienced son. Kinship was also established through marriage. Nicholas Arnold's daughter, Katherine, married William Weston. John Thare's daughter, Katherine, married William Tute, son of a gardener and a future gardener in his own right; a second daughter, Elizabeth, married John Ambler, a journeyman gardener who eventually inherited Thare's holding. William Arnold was godfather to the gardener William Crockford.[3]

These evident links between individual households engendered a climate of co-operation. Young gardeners raised capital by borrowing from their older and richer neighbours and occasionally took joint leases of garden ground. They also received cash in times of

1. C of WAC, F4538; Guildhall Library, MS 3389/2; GLRO, AM/PW/1682/33; Arch Middx Orig.Wills 27 Feb 1684/5 Box 5; Earle, op. cit., pp. 219–29.
2. Earle, op. cit., pp. 208–9,217; C of WAC, F4538.
3. Guildhall Library, MS 3398/2; C of WAC, F 4538; 1049/10/Box 1/3; GLRO, AM/PW/1670/122.

need from the Gardeners' Company, sometimes dispensed by neighbours who were also officials of the Company. Gardeners witnessed each others wills and compiled probate inventories of their neighbours' goods. Such working together was economically, as well as socially, expedient.

Physical geography also encouraged joint action. Access to the narrow lanes which led to the Thames sometimes necessitated crossing another person's land. Wharfage on the Thames was at a premium, and some leases specified that two gardeners had to share a wharf. Drainage, access to water, the repair of roads, paths, walls and the Thames embankment were all matters which affected many gardeners and in which they co-operated. In the chapter on rents, the process of negotiating new leases in 1675 was described. The gardeners clearly benefited from acting as a group against a landlord's inexperienced agent. The Neat Houses developed a reputation as an area supplying high-quality produce. Once established, it was economically more beneficial for the community of gardeners to maintain the reputation by sharing the knowledge of new techniques and supplying each other with plants and seeds rather than directly competing with one another in the market-place.[1]

The Neat House Gardens were within the jurisdiction of the London Gardeners' Company. Like other livery companies, membership was organized hierarchically: ordinary members, known as the yeomanry; liverymen, who gained that distinction by contributing a fine or a silver spoon; and a governing court of assistants of 24 members, two wardens and a master. From the incomplete membership records that date from early in the seventeenth century, it is clear that, from the earliest days, Neat House gardeners were in the Company. George Bayley of the Neat Houses was master in

1. Guildhall Library, MS 3389/2; MS 389/1, fol. 38; GLRO, Arch Middx Wills 20 Jan 1665/6; C of WAC, 1049/3/21/2; 1049/3/30/42; 1049/3/22/2; 1049/3/18/4; 1049/3/18/26; 1049/3/20/14; 1049/10/Box 1/3; 1049/10/Box 43; Stephen Switzer, *Practical Kitchen Gardener*, 1727, p. 177; Richard Bradley, *General Treatise*, 1723, vol. I, p. 157.

1637, and many Neat House gardeners became liverymen in the first half of the century. Almost all the Neat House gardeners in the 1660s and 1670s were members.[1]

In 1661–2, John Weston was one of the two wardens and at least eight of the 24 assistants were Neat House gardeners. Five out of 34 liverymen were from the Neat Houses: Nicholas Arnold, William Grey, John Thare, Edward Boyton, and Richard Wright. In addition, there were eight ordinary members from the Neat Houses, some of whom may have been journeymen. The position was similar in 1671–2 when a quarter of the assistants came from the Neat Houses and a further seven were in the livery. In 1688–9 only three Neat House men were assistants and three more were in the livery, although the master for that year, Robert Chandler, was himself a Neat House gardener. The more well-to-do Neat House gardeners at this time were in the higher ranks of the Gardeners' Company. When one considers the large numbers of gardeners south of the Thames, from Barnes to Greenwich, the growing number of nurserymen in the Kensington area, as well as the market gardeners of all kinds in the old-established gardens in Middlesex to the north and east of the City, the number of Neat House gardeners in the senior ranks of the Gardeners' Company in the second half of the seventeenth century is impressive. Such dominance reflects the importance of the Neat Houses as a source of high-quality produce, the status of the gardeners as technical experts, their wealth, the long time that some of them had been gardeners and members of the Company, and the social cohesion which enabled them to nominate each other to influential positions.[2]

This apparent spirit of co-operation may be modified by the behaviour of some strong-minded individuals, amongst whom was Edward Curtis. He was admitted to the Gardeners' Company when

1. Guildhall Library, MS 3389.1A; MS21.128.
2. Guildhall Library, MS 3389.1A; MS 21.129/1; MS 21.129/4; MS 21.129/8; Thirsk, op. cit., p. 236.

working as an apprentice at the Neat Houses in 1637. He remained in the area for the whole of his career. A liveryman in 1650, he was in the court of assistants by 1661 and elected master for the year 1669–70. He took the inventory of Lott Stalling after Stalling's death in 1684, and he ceased to garden when ejected by the landlord from his holding in 1686 or 1687. Thereafter, he disappears from the records.[1]

If Curtis was fifteen years old when admitted as an apprentice in 1637, he would have been 47 when elected Master and 65 when he ceased gardening. He was, along with the other members of the Gardeners' court of assistants from the Neat Houses in the 1660s and 1670s, one of the older generation. Wisdom did not come with age, for he was obstinate, argumentative, and deceitful. These traits helped him on occasion achieve his own way, but they led ultimately to his downfall. I have already described his opposition to the new rents proposed in 1675: even after he obtained a reduction, he grumbled about the cost of repairs to the river bank bordering his garden.[2]

A few years before this, Curtis had clashed with his colleagues after his year of office as master of the Gardeners' Company. When he stepped down on 10 June, 1670, he was required to render accounts to his successor for all income received on behalf of the Company, less his expenses, and to give up all papers relating to the Company, including a bond from the Blacksmiths' Company for £150 lent to them for three years at interest. Instead, he 'putt off & delayed the same & brought in such layeings out and expenses that the Auditors appointed & swore for the impartial taking thereof could nor would passe the same.' After six months of fruitless negotiations, the Company decided to accept a balance of £8 5s 10d, 'for quietnesse sake & to avoid contests & trouble'. He paid the cash but would not release the bond or other papers, for which the Company

1. Chester CRO, 77/1/39; Guildhall Library, MS 3389.1A; 21.128; GLRO, AM/PI (1)1685/15; C of WAC, 1049/3/5/69.
2. C of WAC, 1049/10/Box 1/3.

sued in June 1671. In his defence, Curtis claimed he really owed the Company only £3 17s 10d. He hoped 'he did honestly discharge' his office as master and would surrender the papers when repaid what he was owed. He denied that he had the bond. No resolution of the legal action is recorded. Curtis's hard bargaining and delaying tactics in reducing the amount to be paid to the Company might just demonstrate him to have been a shrewd businessmen, but the subsequent attempt to go back on his bargain and further reduce his liability hints at dishonesty, a feeling reinforced by the refusal to hand over his accounts and papers and the denial that he held the bond for £150. Had he used the year as master to line his pockets?[1]

Curtis was in severe financial difficulties in 1686 when he owed over £100 in rent to Sir Thomas Grosvenor. He again tried to delay matters as long as possible. Sometime in the summer, Curtis promised to give Grosvenor 'account of the first money that he made of the land til Michaelmas and pay the monies'. He did not do so, neither did he come to some arrangement with Grosvenor when the landlord was in London. In July 1686, Grosvenor lost his patience over the delay which had deprived him of the opportunity to seek another tenant at the earliest opportunity. Instead, he complained, Curtis was 'leaving everything til the last which against the rule of the Bible which is, compound with ye creditor when there is time and not let it go to extremity lest thee be caught and deliver thee to the judge and he send thee to prison.' Grosvenor was still willing to discuss the debt if Curtis quit his land. He did not quit willingly: in the same year, Grosvenor's London agent wrote to a lawyer about Curtis, 'I am ordered By Sr Thomas to Seize a Gardiners Stock and goods and to arrest his person for a great deal of Rent....The Tenants Lease is expired and I desire to know how I shall gett his wife and Servants out of ye house and Land so that another Tenant may come in.' Curtis must have been removed; on 13 April, 1687, Robert

159. PRO, Ham. 481 Gardeners' Co. v Curtis, summarized in Guildhall Library, MS3389/2.

Gascoine signed a new lease for the holding. Curtis's career at the Neat Houses was at a sorry end when, on a small piece of paper dated 20 June 1686, he signed an IOU to Sir Thomas Grosvenor for a featherbed, bolster, pair of sheets, bedcord, hammer and nails: he had to borrow and re-erect his own bed which his landlord had seized. Although other gardeners had arrears of rent at the same time as this episode, and some left their gardens and surrendered their stock to Sir Thomas Grosvenor, none caused so much trouble as the elderly Edward Curtis.[1]

The Neat House gardeners' status was enhanced by their reputation for technical excellence. Although Stephen Switzer still patronized them as 'honest and laborious,' his contemporary, Richard Bradley, accorded individual gardeners the title 'Mr' when writing about them. Two individuals listed in the rate assessment of 1664, John Weston and William Grey, were described as freemen of the City of London, and Lady Scott must have considered Grey's household a suitable place to lodge.[2]

From the information available, we may cautiously conclude that at least the richer members of this group lived like their middle-class neighbours in London and Westminster. In 1747, it was estimated between £100 and £500 was needed to set up as a London kitchen gardener, less than that required to start in many London businesses but still a substantial sum which a successful gardener would expect to enlarge considerably by the time he retired from active business.[3]

1. C of WAC, 1049/3/5/69; 1049/3/5/70; Chester CRO, 77/1/39.
2. C of WAC, F4538; Switzer, *An account of Seeds*, published as part of *The Country Gentleman's Companion*, 1731, p. 50; Bradley, *Philosophical Account*, 1721, vol. I, pp. 108, 157.
3. Earle, op. cit., pp. 32, 36; R. Campbell, *The London Tradesman*, 1747, p. 335.

Figure 23. Covent Garden in 1746. The permanent booths of the larger traders contrast with the baskets of the petty hawkers. A print by J. Maurer.

CHAPTER X
Markets and marketing

'Tom Bold did first begin the Strolling Mart,
And drove about his Turnips in a Cart'

There are studies of the general history of markets and shops in London and Westminster as well as of the market mechanisms of major agricultural produce such as grain and cattle, yet little is known about the way market gardeners, including the Neat House growers, sold their produce. The absence of individual business records leaves unanswered the question of how much control they had once crops left their gardens. For instance, did they take it to market themselves and sell direct to the public or sell it at the market to retailers or wholesalers? Moreover, we can only surmise which markets sold produce from the Neat Houses, or how much was sold by which shops during the latter part of the period under discussion.

These gardens had an advantage over those in many other areas that some customers came to them. The taverns and pleasure gardens served melons and asparagus, and seasonal produce was bought to take away. Pepys took a party on a trip up the Thames on a Sunday in August 1666 and 'at the Neat Houses I landed and bought a millon [melon]'. Richard Bradley implies this practice of direct purchase continued. He recorded in May 1723 that 'Collyflowers, of the right sort were sold in the Gardens for 5s each'. Selling vegetables to visiting gentry might account for only a fraction of total sales but it had the merit that marketing costs were avoided, the produce was as fresh as could be, enhancing the reputation of the gardens for high quality, and the Neat Houses were identified with the vegetables bought, unlike an indeterminate purchase from a market stall.[1]

The bulk of produce went to market. An essential preparation was packing: asparagus with damaged tips, or blemished cauliflowers, would make no show. All vegetables are vulnerable especially if young, early, and of high quality. Wicker baskets were in common use for moving vegetables by the seventeenth century. Pictures of Covent Garden market in the eighteenth century depict a variety of such containers, including the round baskets which, within living memory, porters balanced on their heads in towering piles. Robert Gascoine, the Neat House gardener who died in 1718, possessed a selection of 'maunds and basketts', each for a particular purpose.

Most produce went to market by boat in the seventeenth century. Contemporary pictures of the Thames abound with barges full of all manner of agricultural produce floating eastwards to London. Robert Gascoine possessed '2 boats with Sculls and oares' and it is likely that others were similarly equipped. Whether these boats were their main transport, or small craft for their own use, with market transport being undertaken in larger vessels by boatmen, is unknown. Richard Steele in the *Spectator* painted an idyllic picture of life on the Thames upstream from London in the early morning:

> When we first put off from Shoar, we soon fell in with a Fleet of Gardeners bound for the several Market Ports of London; and it was the most pleasing Scene imaginable to see the Chearfulness with which those industrious People ply'd their Way to a certain Sale of their Goods. The banks on each side are as well peopled, and beautified with as agreeable Plantations, as any Spot on the Earth; but the Thames itself, loaded with the Product of each Shoar, added very much to the Landskip.[2]

1. Massinger, op. cit., p. 33; *The Diary of Samuel Pepys*, *ut sup.*, vol. V, p. 365; Bradley, *General Treatise*, 1723, vol. I, p. 109.
2. Pieter Angillis, 'Covent Garden', *c.*1726, Yale Center for British Art, Paul Mellon Collection; GLRO, AM/PI 1718/18; C of WAC, 1049/10/Box1/3; PRO, PROB3 26/198; Tom Brown, 'Amusements serious and comical', 1700, quoted in Frank Muir, *Oxford Book of Humorous Prose*, 1990, pp. 29–30; *Spectator*, no. 454, Monday August 11, 1712.

Produce not sent by water went overland, by horse and cart or foot-carrier, and that by boat would have to be transhipped to land transport for the final entry into the city markets. Specific references to produce moved from the Neat Houses to market on foot all occur in the late eighteenth and early nineteenth centuries, although large areas of London markets were devoted in the seventeenth century to people who sold vegetables from baskets called 'dorsers', which they carried in on their backs. In early-eighteenth-century pictures of Covent Garden, the majority of vendors are selling from baskets set on the ground (figure 23).[1]

The oldest public market in Westminster was at King Street. The City of London had thirteen markets in 1601, some of them general markets at which vegetables could be sold. They were, in theory, closely regulated and food could generally only be retailed to the public. The markets were held in City streets and became more and more congested in the first half of the century. The Great Fire in 1666 eased this problem: out of the devastation, purpose-built markets were created. The growth of London necessitated the creation of new food markets. Some, like Covent Garden, began before the Fire; many more appeared afterwards. By the end of the seventeenth century, there were at least fifteen suburban markets.[2]

Leadenhall and the Stocks (or Woolchurch) Market were the main post-Fire City markets for vegetables. Of the two, the Stocks was closest to the river and more convenient for water-borne produce. Some of the gardeners' boats mentioned in the *Spectator* in 1712 were heading there. The market, as rebuilt after the Fire, had a few fixed stalls for fruiterers and butchers, but was mostly open space to accommodate the trestle tables, baskets and carts of vegetable sellers for whom it was mainly intended. We do not know if the Neat House

1. Betty R. Masters, *Leybourn's Plans of London Markets, 1677*, London Topographical Society, 1974, p. 35; P.V. McGrath, 'The Marketing of Food, Fodder, and Livestock in the London Area in the Seventeenth Century', (unpublished thesis, Univ. of London, 1948) , pp. 195–6.
2. McGrath, op. cit., pp. 52, 64, 74; Betty R. Masters, op. cit., pp. 13–20.

gardeners used the Stocks to any extent. It was a general market, selling large quantities of peas, beans and turnips in season and may not have been of interest to those selling high-quality vegetables. At least 150 gardeners from Southwark went there in the 1670s, and the market was also frequented by growers from Croydon.[1]

Of the suburban markets which appeared in the seventeenth century, the most likely to interest the people from the Neat Houses were St James's and Covent Garden. St James's, near Jermyn Street, was closest to home but easier to reach by cart than by river. It was noted for meat, fowl, roots and herbs. In 1720, the topographer John Strype thought it 'a market now grown to great account, and much resorted unto, as being well served with good provisions.' It served a well-off area, where people were willing to pay for better vegetables.[2]

Covent Garden was a little to the east of St James's, closer to the City. The market started unofficially in the 1640s in the square or piazza at the heart of the estate developed by the earls of Bedford. A charter was granted to the Bedfords to hold a market in 1670 and by the eighteenth century it had become an important for vegetables, flowers and fruit. The proximity of fashionable West End suburbs led it quickly to specialize in high-quality goods. The mixture of open space, colourful produce and well-clothed gentry made it a frequent subject for artists. In their pictures one can see gentlewomen and their servants strolling through the lines of fruit and vegetable sellers. Fashionable shops in the area for lace, mercery, and clothing, together with taverns, tea and coffee houses, theatres and prostitutes made Covent Garden attractive to those with money and leisure. By 1727, it could be described as 'a market for fruits, herbs, roots and flowers every Tuesday, Thursday and Saturday; which is grown to a considerable account, and well served with

1. Masters, op. cit., pp. 29–37; McGrath, op. cit., pp. 195–6.
2. A.B. Robertson, 'The suburban food market of the eighteenth century', *East London Papers*, vol. 2, no. 1, 1959; John Strype, *Survey of London*, book VI, p. 83; Bradley, *General Treatise*, vol. II, p. 157.

choice goods, which makes it much resorted to.' The *Spectator* commented that its fashionable atmosphere rubbed off on the gardeners supplying the market by barge, 'There was an Air in the Purveyors for *Covent-Garden*, who frequently converse with Morning Rakes, very unlike the seemly Sobriety of those bound for Stocks Market.' The market came to dominate the wholesale trade in best produce in the course of the eighteenth and early nineteenth centuries and held on to much of the retailing, despite competition from shops.[1]

Selling vegetables in early modern London was a more straightforward business than say, selling grain or beef. There were fewer middlemen involved. There is no evidence that forestallers bought crops in the ground or in bulk at the farm gate. There were wholesalers, but they operated in the open markets alongside retailers. Shops were not a major outlet. This simplicity reflects the nature of vegetables: they were not durable; they could not be stored or moved in great bulk without damage, unlike commodities such as grain or malt. These features were especially true of the luxury produce which formed much of the Neat Houses' output.

Another characteristic of London greengrocery was the important part played by women. They predominate as stallholders in depictions of Covent Garden dating from the early eighteenth century (figure 24). Some, like the painting by Pieter Angillis, show sales handled entirely by women, the men acting as porters. Many of the women were small retailers making a precarious living from day to day but some may have operated on a larger scale, either as wholesalers or specialist retailers. Melons were sent by barge to one such businesswoman in 1712, consigned to 'Sarah Sewell and Company, at their Stall in Covent Garden'.[2]

1. Ronald Webber, *Covent Garden: Mud Salad Market*, 1969, pp. 34, 37, 39; *Spectator*, no. 454, Monday August 11, 1712; J.C. Loudon, *Encyclopaedia of Gardening*, 1824, p. 1061.
2. Sir Richard Phillips, *A Morning's Walk from London to Kew*, 1817, pp. 224–6; *Spectator*, ibid.

Figure 24. A quiet corner of Covent Garden Market in 1769. Detail from a study by Thomas Sandby.

The City food markets of the seventeenth century were meant to be retail, for producers to sell direct to consumers. Many gardeners did this. In the 1670s, however, some stallholders in the Stocks sold fruit and vegetables wholesale to retailers from other City markets, mirroring developments at Covent Garden. This trend towards specialism continued through the next century.[1]

Vegetables were also sold by shops and hawkers, though shops, here defined as permanent structures with glass windows and counters, were not important in vegetable retailing in London. Green-cellars or root-cellars, which might be market stalls with storage space underneath, or cellars in buildings rented out to retailers who sold a narrow range of produce, were to be found throughout the period. They predate shops by over a century. In 1597, cellars were said to be a health hazard. A petition called for 'order to be taken for restraint of those that let out cellars and sheds under the stalls where herbs, roots, fruits, bread, and victuals are noisomely kept till they be stale and unwholesome for man's body, and then, mingled with fresh herbs of the same kind, are brought forth into the markets and there sold to the great deceit and hurt of the people.' Both Leadenhall, and the Stocks were reconstructed after the Great Fire with permanent stalls and cellars beneath. Covent Garden also conformed to this pattern in the 1670s. These were not, however, shops selling high-quality produce but simply stalls with storage for roots and cabbages which had a limited life. Such cellars were often mean affairs: in the eighteenth century, 'The stock-in-trade of a green-cellar, for instance, might be "no more than a gallon of sand, two or three birch-brooms and a bunch of turnips".'[2]

The first true greengrocers' shops, selling a range of high quality vegetables, appeared in the next century as a small part of the trans-

1. Betty R Masters, op. cit., p. 13; McGrath, op. cit., pp. 52,74; L.G. Bennett, 'The development and present structure of the horticultural industry of Middlesex and the London region', unpublished Phd thesis, Univ. of Reading, 1950, p. 194.
2. BL, Lansdowne MS. 74, ff.75–6; Masters, op. cit., p. 30; Webber, op. cit., p. 38; M. Dorothy George, op. cit., p. 161.

Figure 25. Hawking asparagus in 1711, by Marcellus Laroon.

formation in middle- and upper-class patterns of consumption. Bradley commented on the high bills that some well-off London families ran up with 'Fruiterers and Herb-shops' in the 1720s. Neat House produce, being high quality or out-of-season, may well have been sold in such places. They were, however, slow to capture the custom of the wealthy from traditional market purveyors.[1]

Costermongers and hawkers retailed vegetables and other goods in London streets for centuries, as woodcuts and prints (figure 25), and musical compositions based on their street-cries, attest. The noise and congestion caused by hawkers selling vegetables, fruit and other food is captured in a poem from 1708:

> Tom Bold did first begin the Strolling Mart,
> And drove about his Turnips in a Cart;
> Sometimes his Wife the Citizens wou'd please
> And from the same machine sell Pecks of Pease.
> Then pippins did in Wheel-barrows abound,
> And Oranges in Whimsey-boards went round.
> Bess Hoy first found it troublesome to bawl,
> And therefore plac'd her Cherries on a Stall;
> Her Currants there and Gooseberries were spread,
> With the enticing Gold of Ginger-bread:
> But Flounders, Sprats and Cucumbers were cry'd
> And every sound and ev'ry Voice was try'd.

These street traders were humble people who could find no other work, like the Irish weaver, born in 1708, who came to London and, 'when his business was dead he sold butter, eggs, roots, greens or any small things he was capable of'. Crops of better quality, such as might come from the Neat Houses, were not, in the first instance, sold to these itinerants. They might end up with them when supplies were plentiful. 'The retailers who keep shops and stands, never buy more than they know they can sell well. And therefore both growers and consumers are much indebted, for the moderate price, and the

1. Bradley, op. cit., vol. I, 1726, p. 150; A. Wilbraham & J.C. Drummond, *The Englishman's Food*, 1957, p. 192.

consequent increased consumption, to the jack-ass drivers, barrow-women, and other itinerant dealers in these articles, who buy of the gardeners in the markets, and hawk through the streets of London, and its environs, vegetables and fruit at a very moderate price.'[1]

1. Richard Dering, *Cries of London,* ed. Denis Stevens, Pennsylvania State Music series, V, 1964, pp. 14–18, 21, 28; [Dr William King], *The Art of Cookery In Imitation of Horace's Art of Poetry*, 1708; Dorothy George, op. cit., p. 190; Middleton, *View of the Agriculture of Middlesex,* 1798, p. 267.

CHAPTER XI
The last years

'That gardening is not altogether an unprofitable activity...we may have ample proof.'

Between the 1720s and the very end of the century little information on the Neat Houses has been found. John Middleton breaks the silence in 1798 writing in his detailed survey of the agriculture of Middlesex of their continued renown for intensive techniques and his respect for their excellence. (It is significant that in writing a book on Middlesex, he chose to describe in some detail these gardens over the county boundary, in Westminster.)

Middleton was impressed by the large sums the gardeners were prepared to pay out to run their business. Rents were £6 to £7 an acre in the 1790s and when Thomas Cubitt acquired the gardens for building in 1825, he had to pay £12 5s an acre on the Grosvenor estate and an average of £10 13s 4d on the Wise estate. Labour costs were estimated at £35 per acre in 1798. The gardens consumed vast amounts of dung; in 1798 they used an estimated sixty cartloads or more per acre. At an average price of five shillings a load (two cartloads), this represented expenditure of £7 10s an acre. One gardener brought 'six hundred loads of dung annually from town', for use 'all on nine acres of ground.' The dung was 'brought from the stables, and shot immediately from the carts in which it is brought, into an oblong heap. To this is daily added, what is brought home in the carts on their return from Town'.[1]

1. C of WAC, 1049/3/5/73; Middleton, op. cit., pp. 259, 263–4, 384, 387–90.

Figure 26. The embankment near the Neat Houses in 1794. From here produce was shipped down the Thames, and those coming by boat seeking pleasure at the gardens might alight. From a watercolour by J. Farington.

Middleton thought dung a major factor in the long success of the Neat Houses:

> This land has been long, or perhaps longer, in the occupation of kitchen-gardeners, than any other land in Britain, and for a great length of time has been supplied with dung, as much in quantity, and as often repeated, as, in the opinion of the occupiers, could be applied with advantage to the crops...one thing they unanimously agree in, namely, that to dung plentifully, and with discretion; to dig the soil well, and to sow good seed; is the only practice on which a reasonable expectation of good and plentiful crops can be founded.[1]

With the aid of dung, the gardeners continued the tradition of intensive cropping. A typical example from 1798 was:

> Soon after Christmas, when the weather is open, they begin by sowing the borders, and then the quarters, with radishes, spinage, onions, and all other seed crops. As soon afterwards as the season will permit, which is generally in February, the same ground is planted with cauliflowers, from the frames, as thick as if no other crop then had possession of the ground. The radishes, &c. are soon sent to market; and when the cauliflowers are so far forward as to be earthed up, sugar loaf cabbages are planted from the aforesaid seed crops. When these are marketed, the stalks are taken up, the ground cleared, and planted with endive and celery from the said seed crop; and daily as these crops are sent to market, the same ground is cropped with celery for winter use.
>
> Thus, by an union of natural fertility with heat (raised by dung), and a degree of moisture, are the occupiers of these grounds enabled to raise the greatest crops in the least possible time.[2]

The main markets for the Neat Houses in the late eighteenth century remained in the cities of Westminster and London although the way the produce reached them had changed. Middleton makes no

1. Middleton, op. cit., pp. 261–3.
2. Middleton, ibid.

specific mention of carriage by water and assumes gardeners will possess horses and carts. Those who surrendered their tenancies in 1825 had packing-sheds, cart-sheds and stables, whereas no livestock at all appears in inventories of their seventeenth-century predecessors. Improved roads (and river crossings) may, by the late eighteenth century, have lessened the advantage of river transport (figure 26). Sir Richard Phillips, in 1817, describes market gardeners travelling overnight in large numbers to ensure produce reached the markets fresh:

> All the roads round London, therefore, are covered with market-carts and waggons during the night, so that they may reach the markets by three, four, or five o'clock, when the dealers attend....This rapid routine creates a prodigious quantity of labour for men, women, and horses. Every gardener has his market-cart or carts, which depart at ten, eleven, twelve, or one o'clock, according to the distance from London.[1]

Much produce was still carried on foot to market, a reflection of the low cost of seasonal, female labour, and of the demand for high quality vegetables which would receive less damage transported in this way. Middleton marvelled at the energy of Welsh women labourers who 'labour several hours in the garden, and go to and from the London markets twice a day, though at from four to seven miles distance.'[2]

In the late eighteenth century many writers attempted to quantify the costs and profits from many types of agriculture and horticulture. These should always be examined sceptically. Some represented the best which could be achieved, most were based on 'normal' expectations, but all simplified matters, often taking no account of depreciation of equipment, the cost of finance or, in some cases, marketing and transport costs, taxes and tithes. They also ignored

1. Middleton, op. cit., p. 264; Sir Richard Phillips, op. cit., pp. 224–6; C of WAC, 1049/10/1/2.
2. Middleton, op. cit., p. 382.

the multitude of risks faced by the kitchen gardener. More things could, and did, go wrong with the intensive production of vegetables than in most branches of agriculture. Back in the seventeenth century, Walter Blith predicted high returns to a kitchen gardener only 'if his commodity prosper well, as some have done; but in the case of non prosperity, some are half undone again, as if it thrive not exceedingly in the growth, prosper not as well in the ripening, escape frost, and thieves, and meet with a good market, what it will come to then I determine not.'[1]

Despite their shortcomings, these attempts at quantification are worth examining, especially any relating to the Neat Houses. Middleton published an assessment of returns per acre, gained by the Neat House gardeners, made by a colleague who lived nearby:

The radishes &c.	£10
Cauliflowers, frequently [£]70 or more but say	[£]60
Cabbages	[£]30
Celery, the first crop, not unfrequently upwards of sixty [pounds], but say	[£]50
Endive	[£]30
Celery, the second crop	[£]40
Total annual produce of one acre	£220

> This he stated as an estimate rather under the mark. Some seasons occasion a considerable loss, perhaps of one crop; but this does not often occur, he was of the opinion that, upon the whole, two hundred pounds an acre was a very low estimate of the average annual produce of these gardens.... The very great expences, in labour, manure, &c. which kitchen-gardeners are at, is evident to every one who lives in the neighbourhood of them. Probably their expences may be thus divided, viz. in labour, 35*l.*; teams and dung, 25*l.*; rent, taxes and tithes, 12*l.*; marketing and expences, 8*l.*; which taken from the foregoing sum of 200*l.* leaves 120*l.* per acre as interest of capital, and profit.[2]

1. Walter Blith, *The English Improver Improved*, 1653, p. 273.
2. Middleton, op. cit., pp. 263–4.

Richard Weston produced a more detailed set of figures to illustrate the economics of multiple-cropping in London kitchen gardens in the 1770s. He envisaged an acre producing, over a year, early cauliflowers under glass and autumn cauliflowers, spinach, lettuces, cucumbers under glass, and endive. The expenses and returns were calculated to be:[1]

	£	s	d
For seed, and raising of 2200 collyflower plants to plant out	1	10	0
Digging 160 rods at 3d per rod	2		
Planting them, hoeing, and earthing them up before winter	1		
Seed, and sowing the ground with spinach and lettuce seed	1	10	
Ten loads of dung for a hot-bed for raising 3300 good cucumber-plants, at 5s per load	2	10	
Digging a row four feet wide between the glasses, two fifths of an acre	0	16	
Seed and labour in raising and transplanting them under the glasses	2		
Earthing up the cucumbers after the collyflowers shall have been gathered	1		
Seed, and raising 7000 good plants of endive	0	14	
Digging a row of ground, two feet and a half wide, between the cucumbers	10		
Planting the endive	0	7	
Seed, and raising 2200 good collyflower plants fit to plant out	1	10	0
Planting them, pulling up the cucumbers in September, earthing them and tying up the endive	2		
Rent 5*l.* Dung, and repairing the glasses, 7*l.* 13s	12	13	0
	£30	0	0

1. Richard Weston, *Tracts on Practical Agriculture and Gardening*, 1773, pp. 56–7.

The produce			
160 rods of spinach, 1s each	8	0	0
9680 coss lettuces 1d	40	6	8
2200 collyflowers 3d	27	10	0
1100 glass cucumbers 6d	27	10	0
7000 plants of endive at 1 farthing	7	15	10
2200 collyflowers 1d	9	3	4
	£120	5	10

For those gardeners who made the profits estimated above, and avoided the many pitfalls of intensive gardening, the rewards were good. Middleton concluded:

> That gardening is not altogether an unprofitable concern (as Dr. Adam Smith has represented it to be), we may have ample proof, both in Middlesex and Surrey. There are generally some gardeners in the commission of the peace. It has produced several sheriffs of counties; and more who have realized from 20 to upwards of 50,000 *l.* in addition to their patrimony.[1]

* * * * * *

The long history of gardening at the Neat Houses came to an end in the second quarter of the nineteenth century. The high rents commanded by horticultural land, coupled with the relatively poor land communications with the rest of the capital, helped shield the area from industrial and domestic building that had occupied land on all sides. Notwithstanding, fifteen acres on the Thames bank had already been lost in 1807 to dock and factory buildings which took advantage of good river transport.

In 1820 the southern portion of the Grosvenor-owned Neat House land, about a quarter of the total owned by the estate and including the 1807 development, was leased to John Johnson. He laid out

1. Middleton, op. cit., p. 268.

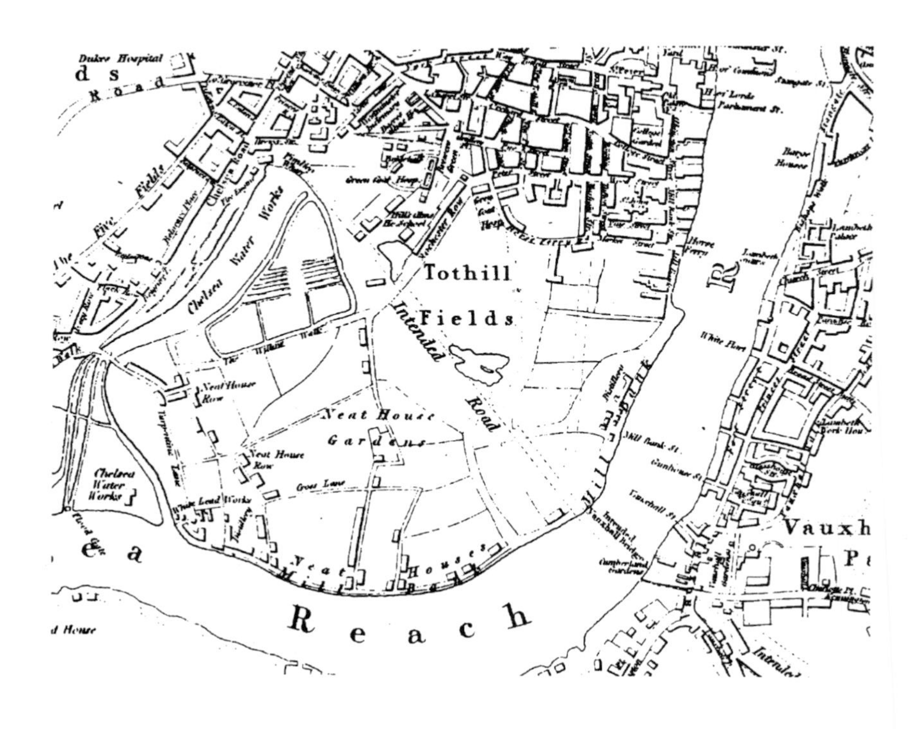

Figure 27. The Neat House gardens in 1813. A white lead works, a distillery and scattered houses already encroach on the gardens. From John Lockie's *Topography of London* (1813).

roads, disturbing gardeners and taking some of their land. In 1825, the builder Thomas Cubitt signalled the end of gardening by leasing all the Grosvenor land, soon afterwards leasing garden land owned by other landlords. Although development took some time, by the end of 1830 Cubitt had dispossessed gardeners of half of their land as roads were laid out. The large-scale development by Cubitt, solving the problem of communications and drainage by a comprehensive infrastructure of roads, sewers, and bridges, made the Neat Houses far more valuable as building land than gardens. It was only a matter of time before gardening was a memory.[1]

To the end, the occupiers displayed a united front in response to external interference. When Johnson started to lay out roads in 1820, he encountered strong opposition from gardeners affected, even though they received land elsewhere in compensation. With wry understatement, the gardeners were said to have been not 'quite cordial to the business'. They had little choice but to give in gracefully when Cubitt took control of the whole area in 1825 but they were handled carefully. Rather than simply tell the gardeners to remove their sheds and outbuildings, Cubitt offered to buy them and suggested to the Grosvenor estate that he should see the gardeners himself to negotiate the sales. He received the surrendered land in person on Christmas Eve 1825. Thus the gardeners ended their association with the estate in a single negotiating group in much the way their predecessors had begun it in 1675.[2]

1. H. Hobhouse, *Thomas Cubitt: Master Builder*, 1971, pp. 167–184; *London And Its Environs*, B.R. Davies, 1841 (map).
2. Hobhouse, op. cit., pp. 168-9, 172.

APPENDIX

THE PROBATE INVENTORY OF ROBERT GASCOINE, TAKEN 27 FEBRUARY 1718.[1]

The first couple of inches of the inventory are largely illegible. This includes goods in the garrets, one or more chambers and the hall and/or parlour. In the damaged part the following entries can be distinguished:

Four pairs of sheets, one chest of drawers, one tea [board?], a fire shovel and tongs, 7 cane chairs, a [?] of hangings, 6 silver hafted knives, 10 [servers?] of plate, one looking glasse, one old clock.

The remaining inventory is as follows:

In the Kitchen.

Item 3 pottage potts, one range with a set of andirons with Doggs and Cheeks one pair of tongs one fender one Crane 3 spitts two tables two joint stooles 5 chairs one clock & case one screen one box iron and heater 6 brass candlesticks one brass tea kettle & Lamp one copper Coffee pott one wind up Jack one Sauce pan 8 pewter dishes 30 pewter plates 6 pewter porringers one pewter bason one warming pan one sword two Musketts one pistoll one Bayonett & pouch one looking glass one glass punch bowl one Glass Jugg some books and other lumber.

1. GLRO, AM/PI 1718/10.

In the back kitchen called the Brewhouse.

Item two coppers with iron work covered with Led on the edge one pair of old Grates with iron pott hookes one gridiron one horse for cloathes two pailes some earthen ware & lumber.

[No totals given.]

In the home ground.

It. 2 banks at the gate by John Hornsby's end and bounding to Mr Edward Amblers planted three rowes wide with Collyfflower plants & ye peice adjoining to the Banke planted with the Same with Some cabbages plants the whole amounting above 1000; the working of the land.

It. 3 beds of collyfflowers plants under lights, & boxes agst. Mr Amblers ground.

It. 1240 whole bellglasses, in all three grounds.

It. 2 banks (one next Mr. Amblers the other Mr Tutes) planted with collyflower plants working the land &c.

It. 2 banks planted with colleyfflower plants & cabbage plants, one next Mr Tutes the other next to & abounding towards the Thames bank & working the land.

It. 3 long beds sowed with reddish seed & carrott seed in a middle quarter with working the quarter &c.

It. one quarter of glass colley flower plants adjoining to the high bank near Mr Tutes with some colwart between ym. & some young lettuce.

It. one quarter of glass colley flower with some colwart agt. the bank where the lights boxes were wintered

It. all boxes and lights in all grounds.

It. one bed of young sparrow grass plants.

It. one quarter of glass colley flower plants with colwarts next to the house.

It. two beds of young sparrow grass for'cd & the dung under the soil beds.

It. 12 hoes 4 dibbers 6 dung forks 2 pickaxes 3 watering potts 4 spades 2 garden reels with lines 3 rakes 5 dung barrows 6 water barrows 3 hatchetts 2 hooks 3 shovells 6 water tubbs, some brewing vessells 100: of matts 3 bushell of reddish seed one sack of rape seed some lettice seed & other seeds with some Maunds & Basketts.

It. One garden Screen, 2 boats with Sculls and oares.

In the Middle Ground.

It. One bank ag. the pale butting Mr Amblers house sowed with reddish & planted with colleyfflower plants with the fruit Trees agt. the pale &c.

It One quarter of Michaelmas onions being 30 rodd or thereabouts adjoining to the aforesaid bank with corn sallett.

It. One Quarter of Spinage planted full crop with cabbidge plants adjoining to the said Quarter of Michaelmas onions fifty rod thereabouts.

It. A Border of Colewarts agt. the pale.

It.One quarter of cabbage plants planted with colewarts between them agt. Mr Amblers fforty rodd.

It. a piece of sparrow grass by the Middle dipping about 18 rodd.

It. one quarter of spinage planted with wide double rowes of cabbage plants containing 50. rodd or so butting & bounding agt. Mr Hornsby's House.

It. In the ground next to the Monster leading to Chelsea 40 rodd of sparrow grass the first yeares planting one piece.

It. 124 rodd of sparrow grass the first yeares planting in one piece being one yeare old.

It. 32 rodds of cutting sparrow grass with colewarts in the alleys.

It. A bank of Colewarts next Mr Jones's meadow.

It a bank next the Willow walk with wide rows of artichokes planted with Colley flower plants working the land &c.

It. 2 beds of young Sparrow grass plants bounding to the Monster.

It. one bank of artichokes stockes adjoining to Mr Gradee

It. One Quarter of artichokes stockes adjoining the aforesaid [...] next Mr Gradee & working the land &c.

[...] 3 beds of afors'd Sparrow grass under boxes & lights in the Cab

It. one bank of Glass colley flowers & plants between them planted with cabbage plants agt. the 3 little tenemnt.

It. one piece of Artichokes Stocks by the Bank agt. Mr. Randalls wall with the working ground &c.

It. one quarter of spinach planted with cabbage plants a side crop agt. Mr Randalls gate being 40 rodds there-abouts.

It. A small piece of colewart & a small piece of Sallary.

It. a laddar

It. a hovell next to the Monster

It. 3 read hedges and cross hedges to them athwart.

It. A Hovell next to the back gate

It. A dung cart with copsyses

It. One quarter of artichoke stocks with working the land adjouning to the aforesaid Qutr of spinage being 45 rodds or thereabouts.

It. One and the bank of Colley flower plants under pieces of glass under Colewarts.

The above mentioned Goods amount to 198. 8. 10.

Index